U0727361

Fashion
Sample House Design

The designers' unique design concept depends on contour, color, material, lighting, layout and accessories

风尚样板房设计

大连理工大学出版社　　　　深圳市艺力文化发展有限公司 编

图书在版编目(CIP)数据

风尚样板房设计：汉英对照 / 深圳市艺力文化发展
有限公司编. — 大连：大连理工大学出版社, 2011.8
　　ISBN 978-7-5611-6308-5

　　Ⅰ.①风… Ⅱ.①深… Ⅲ.①室内装饰设计—图集
Ⅳ.①TU238

中国版本图书馆CIP数据核字（2011）第125280号

出版发行：大连理工大学出版社
　　　　　　（地址：大连市软件园路80号　邮编：116023）
印　　　刷：利丰雅高印刷（深圳）有限公司
幅面尺寸：245mm × 290mm
印　　张：19.5
插　　页：4
出版时间：2011年8月第 1 版
印刷时间：2011年8月第 1 次印刷
责任编辑：袁　斌　张　泓
责任校对：王秀媛
特约编辑：王　宇
装帧设计：洪　辉
文字翻译：王晓林

ISBN 978-7-5611-6308-5
定　　　价：280.00元

电　话：0411-84708842
传　真：0411-84701466
邮　购：0411-84703636
E-mail：designbooks_dutp@yahoo.cn
URL：http://www.dutp.cn

如有质量问题请联系出版中心：（0411）84709043　84709246

PREFACE   序言

Those who can make you enthralled in dreams, those who can transplant their ideas into dreams, those who have the nerve to turn dreams into reality, those are show flat designers. They lead the trends of living fashion. They use the newest materials and latest techniques to search for future living styles. They make their own dreams mirrored in others' reality, indicating the society's consumption tendency and aesthetic level.

The quick development of China's real estate industry in this decade has been one of the top topics. The dream towards ideal residence now has enough reason to come true thanks to the fast development of China's economy. Everyone has his or her own ideal residence: may be it is a house facing the sea with flowers bloom in warm spring; maybe it is a mansion on the top of a mountain with the close view of the stars and moon. Chinese nation is open-minded. They are willing to try and accept diverse living styles, for instance, from Chinese style to Western style, from ancient to modern, from Southeast Asia to Mediterranean, from Jiangnan style to Interlaken style.

Like China's economy, local designers also grows rapidly. From flowing fashions blindly to have their own ideas, from completely copying to selective reference, they continue to challenge themselves by applying new materials and techniques to make their own dreams.

The dream of residence can be traced to ancient times. As a kind of packaging and sales tool for estate agents, the show flat is the symbol that China's real estate steps into the mature level. Its deeper meaning lies in that estate agents are not merely selling houses but also dreams, a living experience that meets consumers' demands, a dream that can be real. Those who can help the estate agent are show flat designers. They are dream makers. The existence of dreams maybe is just a piece of fragment. You will be part of it if you happen to remember it clearly. As for the initial reason, it is the secret that no one will point it out frankly. Isn't it the same for show flats?

Dream is fleeting. No one but show flat designers can walk into and can feel others' dreams. The method of making "dreams" standing still in the form of images and words is really great. When everyone's dreams are collected here, can you tell that if you are living in your own reality? What is the reality? What is your dream?

　　能让人沉迷于其中无法自拔如在梦境的，能让人在梦中植入意识的，拥有做梦的勇气并且能把梦还原成现实的，非样板房的设计师们莫属。他们引领着人们的居住潮流，他们用最新的材料，最先进的技术探索着未来的生活方式，他们把自己的梦照进别人的现实，指引着社会公众的消费心理和审美水平。

　　这十年来中国房地产业快速的发展一直是人们最为关注的焦点话题。在经济高速发展的中国，人们对于居住的梦想更有理由成为现实。每个人都有着不同的居住梦想，或许是面朝大海，春暖花开，或许是麓山之颠，星月掩云。中国的民众包容性也非常强，他们愿意尝试并接受从中式到西式，从现代到传统，从东南亚到地中海、从江南小镇到因特拉肯小镇不同风格形态的居住理念。

　　同中国经济一样，设计师们也在快速的成长。从追随潮流，转化为拥有独立思考能力，敢于突出自我。从无差别的原搬照抄，发展为取长补短，优为我用。他们不断挑战自我，挑战材料和技术，他们在造自己的梦。

　　家的梦想自古有之，而样板房作为一个包装，作为商品房房地产销售的一种手段，它是房地产市场逐渐成熟的标志。更深层的体现在于，房地产不再是简单的买卖房子，而是在卖一个梦想，一个能成为现实的梦想，一种更符合人们要求的生活体验。能帮助地产商实现这目标的非设计师莫属，他们是造梦师。梦的存在是一个过程，或许只是一个片段，如果你记忆深刻，你已经在过程中，至于那个最初的理由，是永远不可能被捕破的秘密。样板房难道不就是这样吗？

　　梦稍纵即逝，除了造梦师别无二人能走进并感知别人的梦想。以图片文字的手段来定格梦想，集结出版的确是个很好的手段。当大家的梦想都汇集于此的时候，你能感知到自己是真正活在自己的现实意识里么？什么是现实？什么是梦？

CONTENTS 目录

LBA Type Sample Villa, Poly International Golf Garden C01,C02

保利国际高尔夫花园C01、C02地块LBA户型别墅样板房

Design Company: Pinki Interior Design Consultancy CO., Ltd.

设计公司：PINKI品伊创意及设计机构&美国IARI刘卫军设计师事务所

Designer: Danfu Liu

设计师：刘卫军

Location: Nanchang
Area: 370m²
Main Material: rome golden-mesh marble, Iceland gray marble, archaize oak floor, rosewood veneer, rustic tile, paint, wallpaper

项目地点：南昌
建筑面积：370m²
主要材料：罗马金网大理石、冰岛灰大理石、仿古橡木地板、花梨饰面板、仿古砖、质感涂料、墙纸

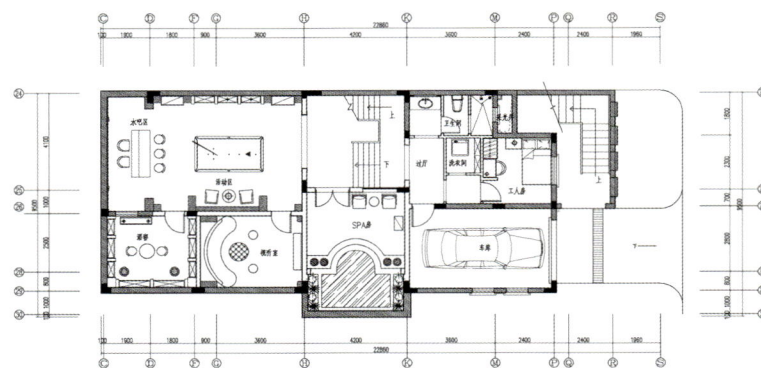

地下一层平面图

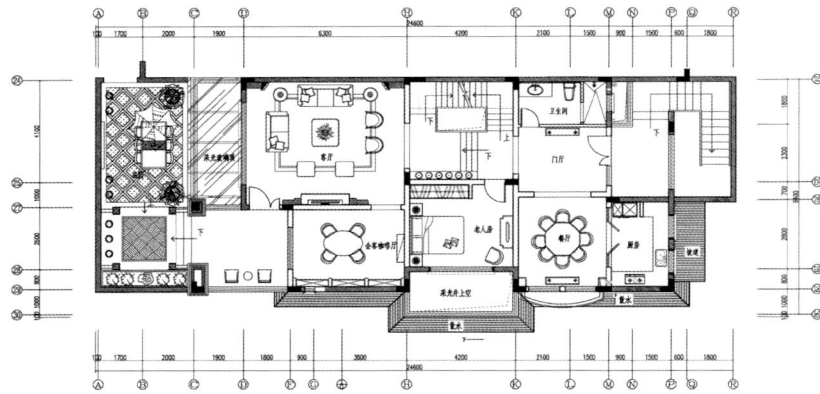

一层平面图

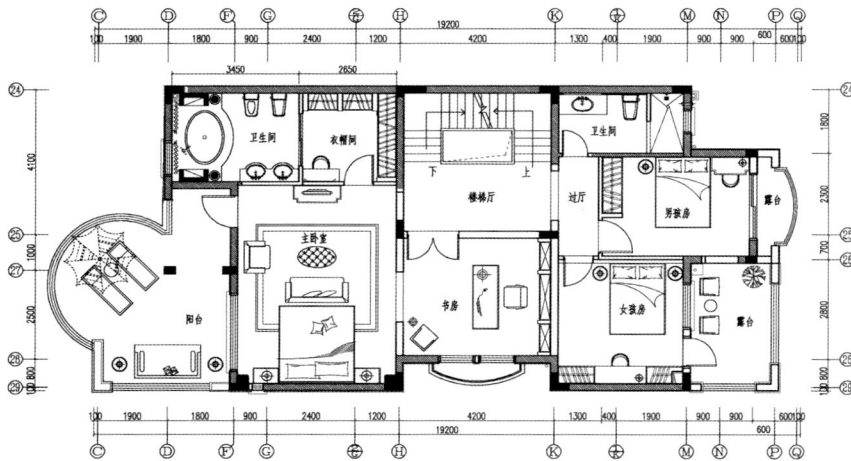

二层平面图

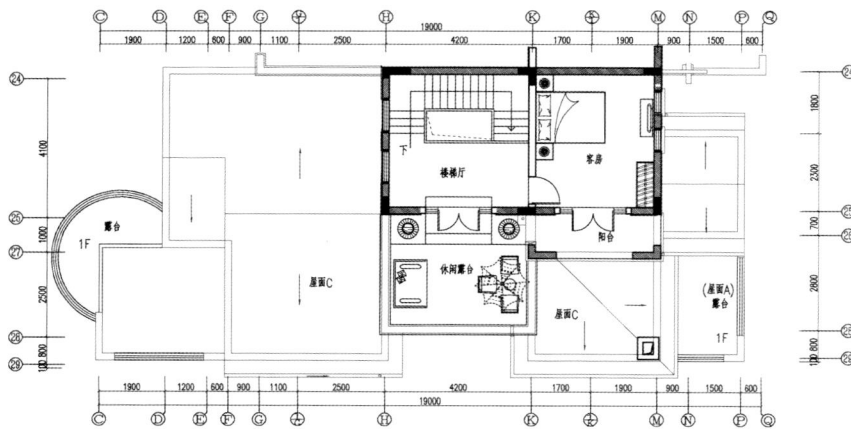

三层平面图

This case with "blossming on the path" as the theme, pursues classical poetry romantic tracks, and induges in the outstanding charm of the aesthtic love. Travelling in history is like walking in the rose garden, you can see those beautiful verse which are as bright as the flowers blooming in the living room, revealing a noble and warm household atmosphere which will not be submerged by the history of the past and the bumpy world.

The mood of ramble on the path is poetic, elegant and quiet, but not in a rush. The flowers are natural, simple and peaceful, but not worldliness. The leisure and romantic in the space make the flowers resplendent, and make people delightful. The blossming on the path in March will be loved by people and drive people crazy, as if the shadow and the charm of flowers are deeply settled in one's body.

　　本案以"陌上花开"为主题，追寻古典诗词的浪漫足迹。沉醉于唯美爱情的旷世韵味。在历史中游历，犹如行走在蔷薇园中，你可以看到那些华美的诗句，一如客厅中绽放的花般鲜艳，处处流露着王朝贵族温馨的家居氛围，不会被历史的沧桑、人世的坎坷所淹没。

　　漫步陌上，心情是诗意的，优雅而散淡，不惹匆促；花是自然的，朴素而恬淡，不落尘俗，如同空间中的那份悠闲与浪漫，花在其中生命得以璀璨，人在其中心情得以畅然。三月陌上花，让人爱让人痴，恍惚人的骨子里头都沉淀了花的影子，花的风韵。

Royal Residence

皇家公馆

Design Company: Shenzhen Wang Wuping Design Studio

设计公司：深圳五平设计机构

Designer: Wang Wuping

设计师：王五平

Location: Dongguan
Area: 170m²
Main Material: Dulux paint, ecological wood, wallpaper, imitational marble tile, art glass

项目地点：东莞
建筑面积：170m²
主要材料：多乐士乳胶漆、生态木、墙纸、仿大理石磁砖、艺术玻璃

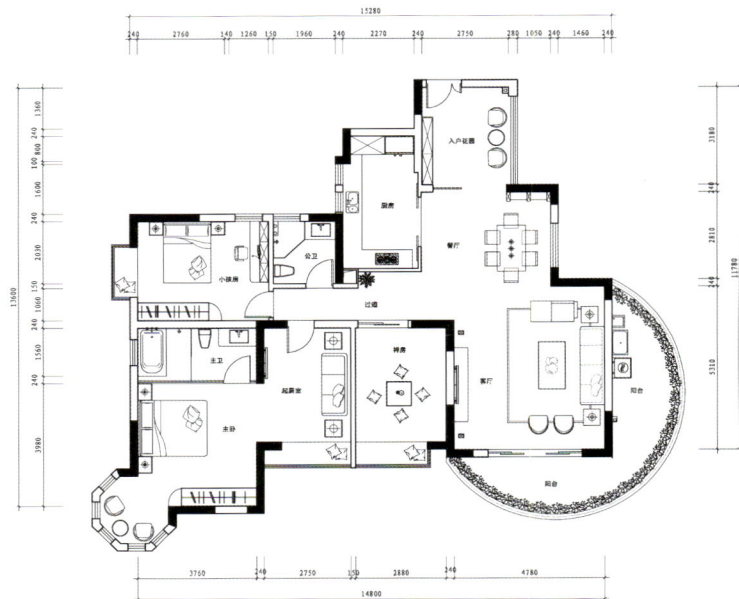

平面图

Bright red and pure white, with the mottled shadow outside the window show a kind of warm and fragrant. And the dreamy blue outside the window is mysterious and wonderful. Such a home, always inadvertently making people feel like in a fragrant, quiet and artistic space. From the space planning, the design combines the storeroom with the kitchen, making a large kitchen feeling. The door of the kitchen is also very creative designed. Breaking the feeling of traditional sliding doors, the designer uses art glass partition as the sliding door of the kitchen, without designing a door pocket. When the door is open, the opening is large, so the kitchen and the living room can be interacted. And when the door is closed, it can be used as a art screen, and can also close the cooking fumes, as while as making a transparent visual effect. The master bedroom is connected with a small room which functions as living room. The wall of master bathroom uses tree-shaped art glass; the glass keeps the broad vision and adds a fashion atmosphere to the bedroom. The background wall of the sofa uses ecological wood of red and black colors, which makes a distinct contrast with the white imitational marble tile, enhancing the visual appeal of the space.

　　鲜艳的红色，纯净的白色，间或有着窗外斑驳阳光的影，有种午后的温馨，而窗外梦幻般的蓝色，神秘而又精彩。这样的家，总是不经意间让人有种香暖入怀，替入梦，淡淡幽意，醉心归的意境。从空间规划上，将厨房里面一个储物间打通，形成一个大厨房的感觉。厨房门也设计得比较有新意，打破以前常规推拉门设计，运用感觉，设计师运用艺术玻璃隔断来做成厨房的推拉门，没有门套。拉开厨房两扇门，门洞比较大，厨房和餐厅空间就可以相互借用；关起来既可以当艺术屏风，还可以隔油烟，又使视觉通透。主卧打通了一个小房间，做成起居室。主卫的墙面设计成树形玻璃，保持洗手间的视觉开阔，更增加了卧室的时尚氛围。沙发背景运用了红黑相间的生态木，和旁边的白色仿石材砖形成鲜明的红白对比，更增强了视觉感染力。

Shenzhen Chunhua Seasons Garden Show Flat

深圳市春华四季园样板房

Design Company: Gu Yanping Interior Design Studio

设计公司：顾彦平室内设计工作室

Designer: Gu Yanping

设计师：顾彦平

Location: Shenzhen
Area: 248m²
Main Material: marble, ceramic stone, cat's eye stone
stainless steel, clear mirror, art mosaic, acrylic sheet

项目地点：深圳
建筑面积：248m²
主要材料：大理石、微晶石、猫眼皮、不锈钢、清镜、艺术马赛克、亚克力板

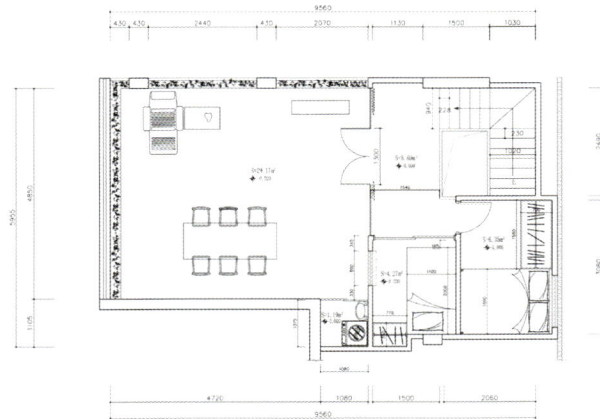

天台平面图

一层平面图

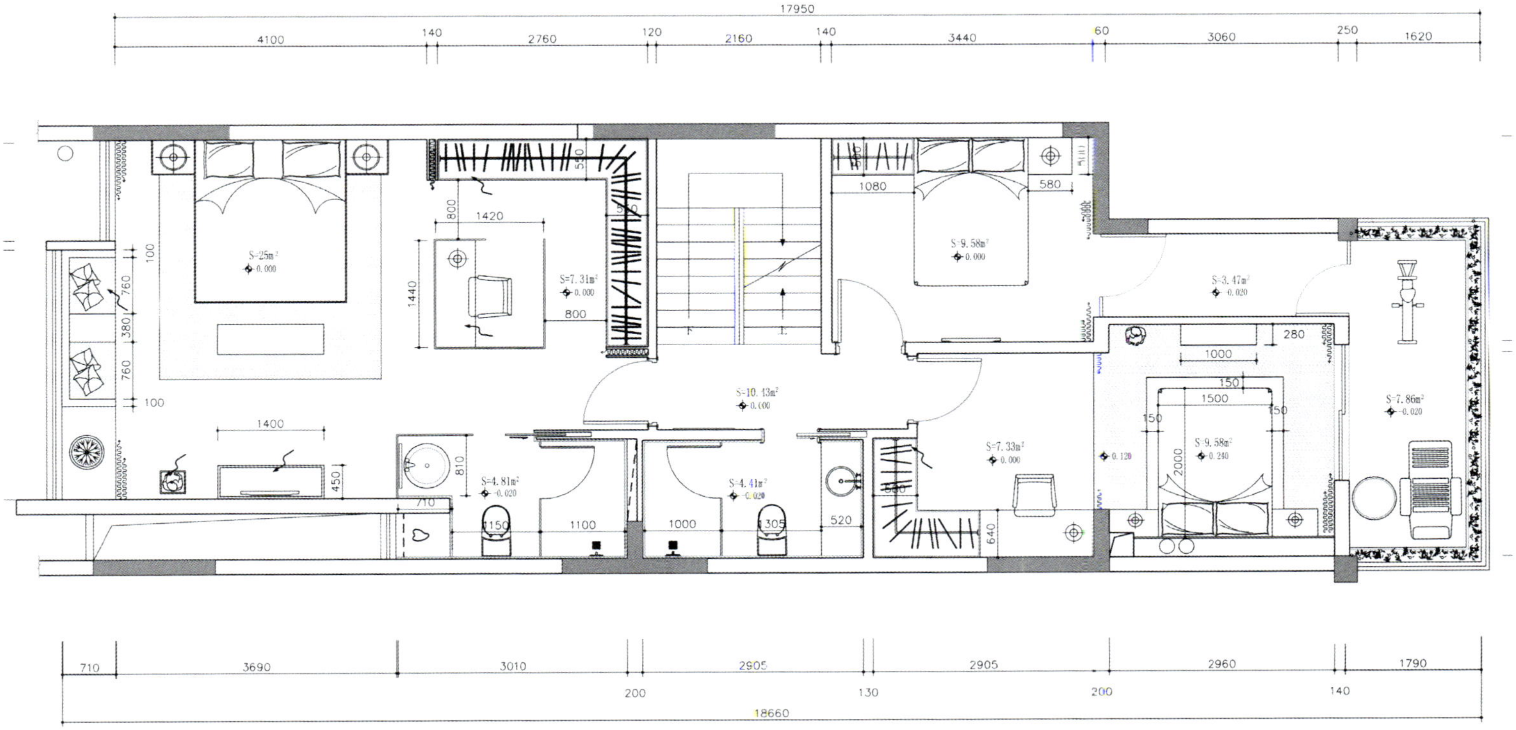

二层平面图

17950

4100 | 140 | 2760 | 120 | 2160 | 140 | 3440 | 60 | 3060 | 250 | 1620

S=25m²
◆±0.000

S=7.31m²
◆±0.000

S=9.58m²
◆±0.000

S=3.47m²
◆-0.020

S=10.13m²
◆±0.000

S=4.81m²
◆-0.020

S=4.41m²
◆-0.020

S=7.33m²
◆±0.000

S=9.58m²
◆-0.240

S=7.86m²
◆-0.020

800 | 1420 | 1440 | 800

760 | 380 | 760 | 100 | 100

1400 | 450 | 710

1080 | 580

1000 | 280

1000 | 150 | 1500 | 150 | 2000

1100 | 1000 | 305 | 520 | 640

710 | 3690 | 3010 | 2905 | 2905 | 2960 | 1790

200 | 130 | 200 | 140

18660

Distinct avant-garde design, dreamy decorative colors and fashionable quality, and the living art of cool and luxury are the ideal choice of urban youth. The contemporary avant-garde style has been the choice of artistic people's residential space, which is individualistic and more expressive than minimalism. A breakthrough of ordinary space structure, bold contrast on color arrangement and balanced material selection and match, combine to seek for a surrealistic balance in coolness. And this balance is no doubt a powerful weapon to against monotonous aesthetics, monotonous living idea and monotonous lifestyle.

In this project, the cool black, white and grey colors are toned with red, which expresses the simple fashion in its overall style. The glass mirror and the stainless steel reflect the cool red colors, flavored with the ultra-modern cat's eye stone to show its uniqueness and noble quality. This cool design style filled with chic elements, is the perfect mixture of ornamental nature and practical function. It not only exemplifies high-end design quality but also brings better and cooler visual experience.

　　个性鲜明的前卫设计风格，梦幻的装饰色彩，定位时尚，缤纷，提倡炫酷奢享的生活艺术是城市年轻族群的理想之选。比简约更加凸显自我、张扬个性的现代前卫风格已经成为艺术人在家居设计中的首选。打破常规的空间结构，大胆鲜明、对比强烈的色彩布置以及刚柔并济的选材搭配，无不让人在冷峻中寻求到一种超现实的平衡，而这种平衡无疑也是对审美单一、居住理念单一、生活方式单一的最有力的抨击。

　　本案设计采用了以炫酷的黑白灰主色调搭配红色色调，整体风格简约时尚。玻璃镜面不锈钢与炫酷的红色相映生辉，加上极富时尚感的猫眼材料，更加显出与众不同的尊贵气质。充满时尚元素的炫酷设计风格追求观赏性与实用性的完美结合，更好地体现了设计的卓越品质，同时也带来更加炫酷的视觉体验！

Jin Hui Yuan

金辉园

Design Company: Kern & Roy Urban Planning and Landscape Design Ltd.

设计公司：柯恩罗伊景观设计事务所构

Designer: Xie Yihui

设计师：谢一晖

Location: Xiamen
Area: 260m²
Main Material: cement paint, laminate floor, rustic tile, glass

项目地点：厦门
建筑面积：260m²
主要材料：水泥漆、金刚板、仿古砖、玻璃

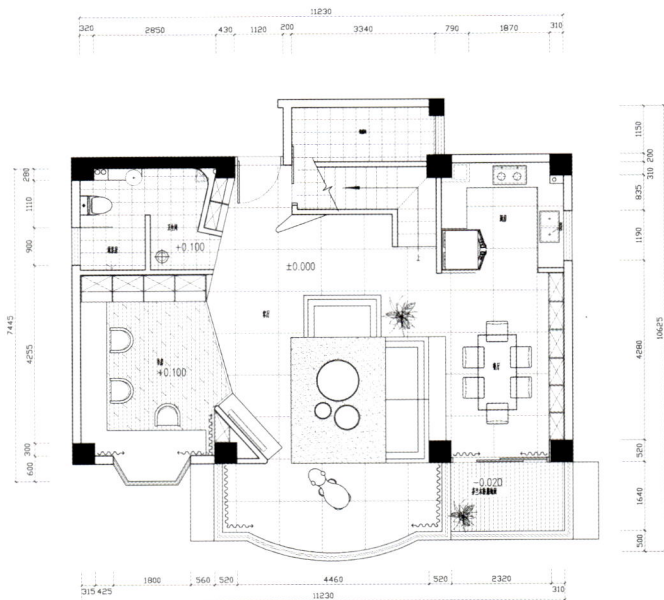

一层平面图

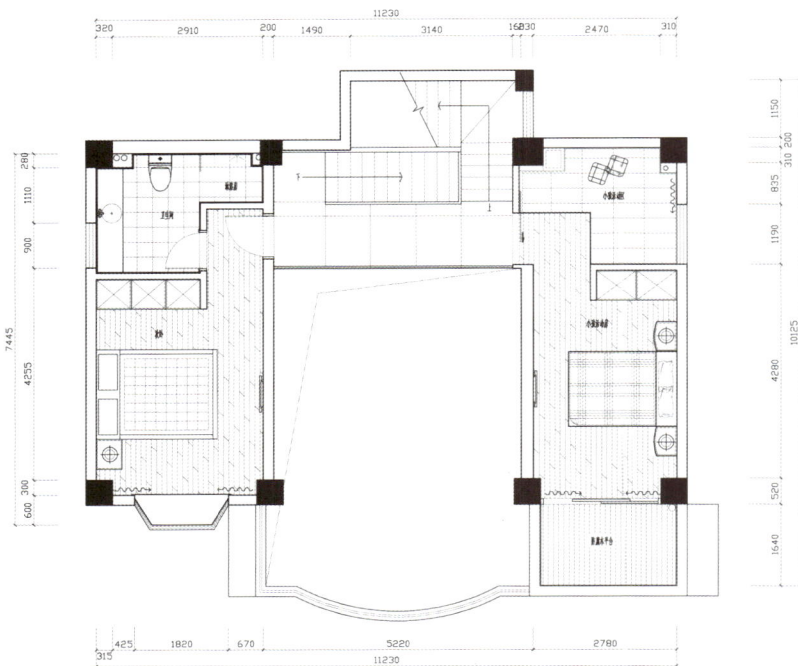

二层平面图

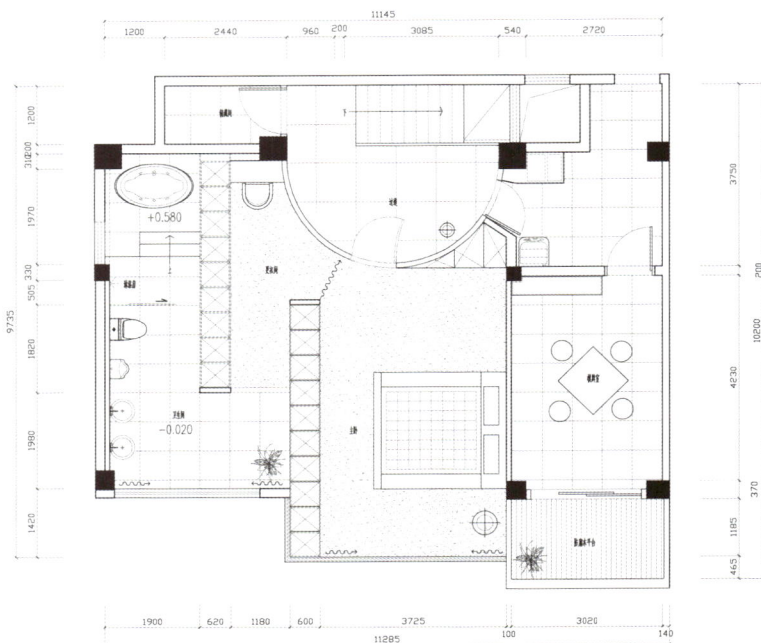

三层平面图

Case is named as "thank you", which shows gratitude to the owner who trust the designer and hand the lovely house to him for designing.

本案取名为"谢谢",是对于业主信任地将她心爱的房子交给设计师设计表示感谢。

Huasheng Hotel Apartment Sample House 13#

华盛领寓酒店公寓样板房13#

Design Company: Eric Tai Design Co. LTD

设计公司：戴勇室内设计师事务所

Designer: Eric Tai

设计师：戴勇

Location: Shenzhen
Area: 55m²
Main Material: clear mirror, grey mirror, marble, wall paper, carpet

项目地点：深圳
建筑面积：55m²
主要材料：清镜、灰镜、大理石、墙纸、地毯

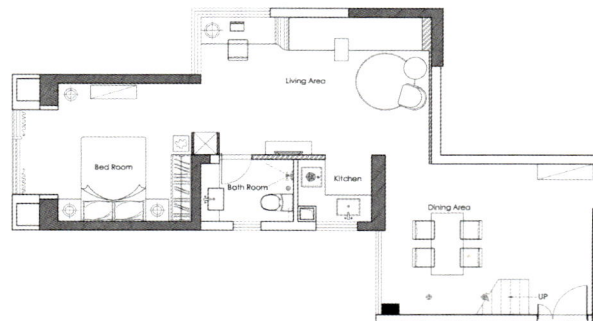

平面图

According to the layout feature of this sample house, the design tries to use liner elements to decorate the whole space, showing the whole space with deep interest of extraordinary personality. Especially in the sitting room, it contains the most abundant linear language, such as the linear volume, the linear surface and then the line itself which express the vivid linear feeling.

Line, the simple geometric language, will thus be actively organized. All the elements are extended around "line": horizontal lines pursuit the visual effect of depth, presenting the tree form. Vertical lines terminate the horizontal frowsy; curved lines appear as active elements. Space adornment language is simple and unified, extremely personalized. The added attic sends the possibility of space transformation, breaking the original layout weakness through innovative and simple space design.

Using concise and personalized element brings strong visual sense, giving a new visual impact to the owner entering in. The layout emphasizing the clear and neat methods, matches with plain and cool style, giving a excellent stage to the dynamic design language. In the meantime, it also conveys a positive and optimistic attitude of life.

依循该公寓的户型特点，设计欲以相吻合的线性元素铺陈整个空间，呈现出与众不同的个性与深深的趣味。特别是客厅，容纳了最丰富的线性语言，条的体块，条的面块，再至条的线，令人目不暇接，表现出灵活生动的线性感觉。

线这一简单的几何语言便由此被积极地组织起来，所有的元素都在围绕"线"而扩展：横的线追求纵深的视觉效果，呈现自由舒展的形态；竖的线，终止横的闷；曲的线，作为活跃的分子而出现。空间的装饰语言单纯而统一，极具个性。增加的阁楼，为主人传递空间改造的可能性，更通过创新简约的空间规划，突破了原有户型的弱势。

设计利用简洁而富个性的元素带给视觉强烈的冲击感，呈现给推门而入的主人一种崭新的空间视觉效果。平面布局干脆利索的手法配以素冷的格调，给予动感十足的设计语言卓越的表演场，同时亦传达出一种积极乐观以及桀骜不拘的生活态度。

Arthur Residence

亚瑟公馆

Design Company: Shenzhen Wang Wuping Design Studio

设计公司：深圳五平设计机构

Designer: Wang Wuping

设计师：王五平

Location: Huadu, Guangzhou
Area: 160m²
Main Material: Dulux paint, oak floor, wallpaper, imitational marble tile

项目地点：广州花都
建筑面积：160m²
主要材料：多乐士乳胶漆、橡木地板、墙纸、仿大理石磁砖

平面图

In this plan, due to the conjoined design of the living room and restaurant, sofa background and restaurant wall is big and long. The designer designs three white arcs which creatively and naturally separate the living and dining areas, and also increase the fluidity and the sense of design. Because the original passage is narrow, the designer broadens the passage through the changing of plane, making a more fluent line between living and dining area and bedroom. Through the wet and dry area design of the public washroom, the washbasin is designed outside and wooden pillars are designed as partition, which broaden the passage and make the passage space more transparent. The wooden pillars also increase the sense of art of the passage. The wall of cloakroom in the master bedroom is designed with art glass, which allows more sunlight in the checkroom, as well as adds the artistic atmosphere in the bedroom.

　　由于本案客餐厅是连体空间，在这样的户型中，沙发背景和餐厅那面墙就显得大而长。设计师在餐厅区域设计三个白色弧形体，极富创意地把客餐厅区域自然分开，另外也增加了空间的流畅感和设计感。由于原户型过道比较窄，设计师通过平面的改动，拓宽了过道，使客餐厅与卧室之间的贯穿动线比较流畅。而且公共卫空间通过干湿分开的功能设计，把洗手台放在外面，设计成几根通透的木柱做隔断。把公共卫空间也无形借给了过道，使过道空间更具有通透性，几根木柱也增加了过道空间艺术形态的生动性。主卧衣帽间那面墙，设计师设计成艺术玻璃，既可以帮助衣帽间照射到自然光，也为卧室增添不少的艺术氛围。

Youngor Future City
雅戈尔未来城

Design Company: WILLIS Design Studio
设计公司：巫小伟设计事务所
Designer: Wu Xiaowei
设计师：巫小伟

Location: Suzhou
Area: 140m²
Main Material: floor, glass, carved decoration, artificial stone countertop, texture tile

项目地点：苏州
建筑面积：140m²
主要材料：地板、玻璃、装饰雕花、人造石台面、皮纹砖

平面图

This case is Younger Future City apartment with three rooms, one dining room, one sitting room and two bathrooms, located in contemporary and concise style, with wood color as the fundamental key. The biggest characteristic is to use wood color veneer to decorate wall and ceiling. The authentic texture ambiguously expressed modern people's back-to-basics pursuit and yearning for nature. Every different functional area is partitioned or continuous, guaranteeing the communication, again not interfering with each other.

The sitting room is divided into reception and repast area, distinguishing through different decoration on the top and changes of ground flooring. The sitting room TV wall is more interesting, using glass and soft decoration, glittering and translucent with rich levels. The space of dining room is narrower, the designer places wine ark made of glass and stainless steel on one side to meet the dining function while extended the space visually. On one side of the study, the designer utilizes the outside as a sun room for the activities or rest of the owner. The design of master bedroom is very simple, using one side of the space as a enter type cloakroom. The ceiling uses traditional plasterboard, and the wall uses wood color veneer together with pieces of decorative pictures, they complete the shaping of space. Second bedroom appears quite lively, black and white lines matching with pink bedding, elegant and moving.

　　本案为雅戈尔未来城三室二厅二卫的平层公寓，定位于现代简约风格，以原木色为基调，最大的特色是采用原木色的饰面板装饰墙面和顶部，原汁原味的纹理隐约表达了现代人返璞归真的追求和对自然的向往。在各个不同的功能区域之间或隔断或隔而不断，保证了空间的交流，又互不干扰。

　　客厅被分为会客区和就餐区，通过顶部的不同装饰和地面铺设的变化区分开来。客厅电视墙比较有意思，采用玻璃和软包相结合，既晶莹剔透又富有层次感。餐厅的空间比较狭小，设计师在一侧安置了玻璃与不锈钢结合的装饰酒柜，满足了餐厅功能的同时又在视觉上拓展了空间。在书房的一侧，设计师利用外部的空间做成一个阳光房，便于业主活动休息使用。主卧的设计非常简单，利用一侧的空间做成一个步入式的衣帽间，顶部采用传统的石膏板吊顶，墙面使用原木色饰面板，挂上几幅装饰画即完成了空间的塑造。次卧则显得比较活泼，黑白色为主的线条配以粉色的床品，清丽动人。

B-1 Typed Model House, 6#, Yuhu Bay

玉湖湾6栋B-1户型样板房

Design Company: Eric Tai Design Co.,LTD

设计公司：戴勇室内设计师事务所

Designer: Eric Tai

设计师：戴勇

Location: Shenzhen
Area: 89m²
Main Material: antique brick, mosaic,texture paint, ocean related wallpaper, vintage oak wood veneer, vintage oak wood floor

项目地点：深圳
建筑面积：89m²
主要材料：仿古砖、马赛克、肌理漆、海洋题材墙纸、橡木作旧木饰面、橡木作旧木地板

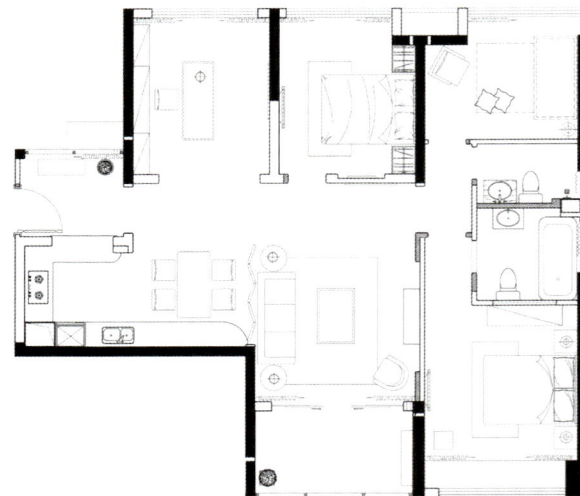

平面图

Blue and white are the two predominant colors of Aegean Sea which is also the basic colors match used in this show flat. The designer firstly applied the pure white and antique beige in the living room while blue is used as decoration in the space or as partial colors complement. The beige wall with the feel of textured lacquer gives a natural impression of Ancient Greek pillars that stand on the sea shore cliff. The archaized white wood veneer and the sophisticatedly-arranged archaized bricks provide a sense of nostalgia. Dark blue bricks are used as the floor of front garden, bluish violet flowers and marine blue glass vase are placed on the dining table, cobalt blue candles are put on top of the fireplace and sky-blue wall paper is used as the background of living room—different shades of blue arranged by the designer appear as petals form a cornflower.

As for the shapes, more rounded corners and curves are employed in order to make the space softer, such as the arched door way, and the bookshelf in the study. Almost all the exposed corners within the dwelling were rounded. The utilizing of doorways improves spatial levels and makes each space a single unit and forms background for other spaces. Iron decorations and lightings, such as the wall lamp and the chandelier in the living room and the iron grate, combined with the rattan furniture and flax carpet, interpret a sense of vintage. The modern-styled mirror hanging above the fireplace injects the space with some cosmo air which lessens the dullness of too much antique and meanwhile maintains the poetic atmosphere that antique provides. Entering the master's bedroom, one can see a spread of artistic and vintage wall paper seeming as if it was washed by sea water which again evokes the visitor's nostalgic feeling. Inky wintry-branch-like iron decoration is the focal point. The delicate lamp with special shape looks like plants growing undersea but now transplanted at the sides of the bed. The design of the washroom follows the overall elegant vintage style of the interior design by using dark blue archaized bricks and mosaic toned with beige textured lacquer.

Here, each room is so aesthetically enchanting that visitors feel as carried away by the exotic ambience. Here, the designer makes great effort to explore the ancient culture from a modern man's perspective and interpret the lifestyle of Aegean Sea by modern design skills. Here, far away from the hullabaloo, as if resting by the Aegean Sea and feeling the wind whispering to you, one feels utterly refreshed by the romance that fills the house.

蓝和白永远是爱琴海的主打色。这套样板房在色彩搭配上也不例外。设计师首先用纯净的白色和古旧的米黄色来铺开整个客厅，而蓝色则多作为空间的点缀色或是局部配色来运用。米白的墙面带着肌理漆的质感，那种天然的感觉好像来自于海边峭壁上屹立的古希腊石柱。白色作旧的木饰面以及地面精心排列的仿古砖，带着斑驳的怀旧感。入户花园地面嵌着深蓝色的地砖，餐桌上摆放着蓝紫色的精美插花，海蓝色的玻璃瓶，壁炉上是钴蓝色的烛台，而客房背景墙上的壁纸则接近于天蓝色。这些空间中出现的蓝，被设计师划分作了很多层次，如同矢车菊的花瓣，深深浅浅。

设计在造型上用到了比较多的圆角和曲线，柔化了空间。如天花吊顶的转角，拱形的门洞，书房的书架等。室内几乎所有的阳角也都作了圆角处理。门洞加深了空间的层次感，使每一个空间既自成一统，同时又成为另一个空间的借景。后期软装上选择了比较多的铁艺饰品和灯具，如客厅的壁灯、吊灯以及壁炉前的铁艺围栏，再搭配上藤编的家具和亚麻地毯，古朴的气息迎面而来。壁炉上那面颇具现代感的镜子又给空间注入了些许的时尚气息，缓和了古旧所带来的沉闷，却又保留下了古老所带给人们的诗意。走进主人房，床头那一片写满怀旧感觉的艺术墙纸仿佛被海水冲刷过一般，再一次唤起人们对古老的向往。枯枝状的黑色铁艺挂饰无疑是点睛之笔。精致而独特的台灯就像生长在海底的植物，被移植到了人们枕边。洗手间的设计同样沿承了整体的风格特色。深蓝的仿古砖和马赛克，搭配上米白色的肌理漆，一样的古朴和优美。

在这里，每一个房间都那么唯美脱俗，让人徘徊在异国情调之中，一再地忘了归路。在这里，设计师仍然在以现代人的身份解读一个古老的文明，用现代设计手法演绎一段爱琴海风情。在这里，没有了尘世的喧嚣，如同栖居在爱琴海边，只有从海上吹来的风在耳畔呢喃，带来清新一室，浪漫满屋。

#E, Block A, South Tower of Hengyuntai Jubao Huafu Model House

恒运泰聚宝华府南塔A座E户型样板房

Design Company: Eric Tai Design Co.,LTD
设计公司：戴勇室内设计师事务所
Designer: Eric Tai
设计师：戴勇

Location: Shenzhen
Area: 73m²
Main Material: Sally Anna marble, sand grey mirror, tea-colored mirror, wallpaper, cream-colored carpet

项目地点：深圳
建筑面积：73m²
主要材料：莎安娜大理石、药水砂灰镜、茶镜、墙纸、米色地毯

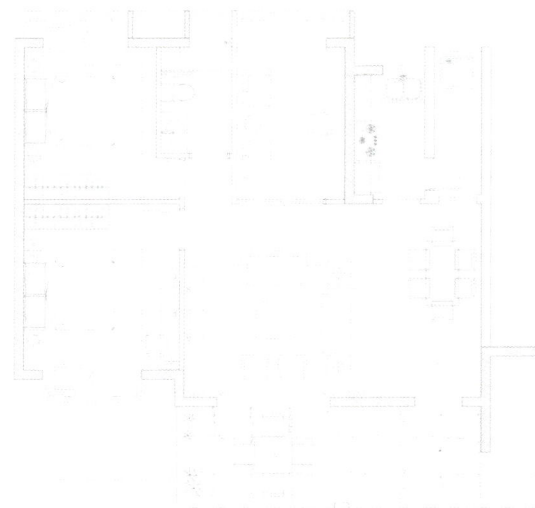

平面图

This model house is the "left bank" of the owner, elegant and noble, describing the unique artistic taste in the calm pace. The integrated living and dining area looks more harmonious due to the silver gloss wallpaoer. The light flower pattern seems like the gentle feeling of the hostess, and the round back steel dining chairs with the same flower pattern blend into this situation, together with the black crystal rectangle dining table constructing a classic group of scenery. The emotion and temperament of the space closely links different functional space. And the selected furnishings, such as the " barrel " shape lamp, extend all the way from the living room to the master bedroom, suspending by both sides of the bed. The silver surfaces make the dark red soft decoration with leather of the head of the bed stand out, charming and fashionable. Luxuriant sense revealed at every incn of the interior decoration, abstract paintings with golden edge and black background, bedside cupboard tables, chairs, and black silk embroidery hold pillow attracting people to feel with hearts and to touch with hands.

Condensed European classical style and luxury fashionable essence deduce a life attitude, and a way of life, graceful and remarkable.

　　此间住宅是主人的"左岸"，优雅而又散发着贵胄气质，在平静的步调中叙述独到的艺术位品。已然一体化处理的客餐厅因为银质光泽的壁纸而更为融洽。淡淡的花朵图案像女主人柔软的情怀以同样的色泽与相似的花朵图案融入这样的情境，透过黑晶长方餐桌构成经典的组景。空间的情感、空间的气质紧密联系着各居一方的功能区域。还有精心挑选的陈设配饰，如"桶"状的灯饰，从客厅一路延续而来，悬挂于主人床的两边，仍旧使用银质的外表将暗红软包皮革床头衬托得娇美时尚。华丽的质感展露于室内每寸装扮、金边黑底的抽象画作、床边柜、桌椅、黑丝刺绣抱枕，吸引人们用心灵去感悟、用双手去触摸。

　　设计凝聚欧式古典与奢华时尚的精髓，演绎出一种生活态度、一种生活方式，尔雅不凡。

B-1 Typed Model House, 3#, Yuhu Bay

玉湖湾3栋B-1户型样板房

Design Company: Eric Tai Design Co.,LTD

设计公司：戴勇室内设计师事务所

Designer: Eric Tai

设计师：戴勇

Location: Shenzhen
Area: 130m²
Main Material: ebony, white dolomite, cream-colored dolomite, silk hand-painted wallpaper, mosaic, walnut wood floor

项目地点：深圳
建筑面积：130m²
主要材料：黑檀木、中花白云石、米黄云石、真丝手绘壁纸、马赛克、胡桃木地板

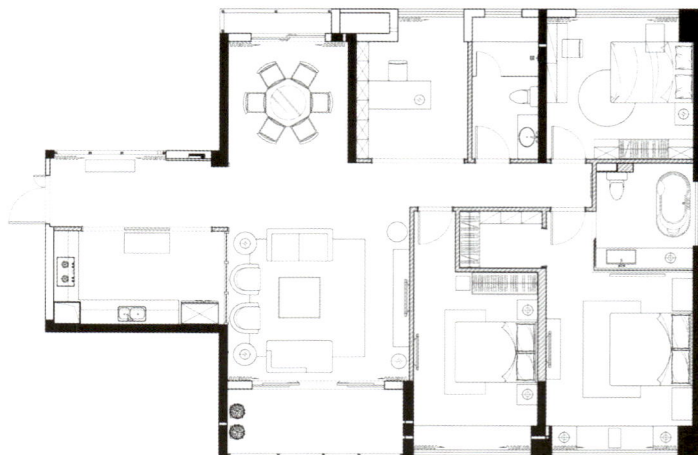

平面图

This is a traditional Chinese style show flat based in the center area of Baoan, Shenzhen. Classic elements were blended in modern design method. The design used modern languages to demonstrate the connotation of oriental culture.

The space was reconstructed. The wall between the sitting room and kitchen was knocked through. The wall of the study was also knocked through to make the study an open space. The glass wall between sofas and the Chinese screen in the kitchen have the function to separate space and to enable space gradation. Chinese elements can be seen everywhere. The delicate woodcarving, the old-fashioned wooded armchair, the elegant Chinese floor lamp, and the elaborated teapot and cups, etc. each item reveals the old tradition style. The materials are all very traditional, such as the off-white marble floor and ebony. Classic Chinese wallpaper and simplified white ceiling is the main color. Red and green acts as decorative role. The designer used modern materials to convey traditional oriental flavors through modern design method.

When you walk around the washroom and enter the master bedroom, the first thing come into view is the customized flower&bird screen. The lantern cover has maples on it. The red bed cabinet is the focal point. The owner can have a good rest after a day's busy work by lying in bed and holding the pillow and having a sweet dream.

　　坐落于宝安中心区白玉湖湾样板房是一套由设计师精心打造的新中式风格居室。经典的东方元素融于当代的设计手法中，寓古于今，设计师用一种现代的设计语言诠释着东方文化的意蕴。

　　该项目在平面上对原建筑布局重新做了调整。厨房与玄关、客厅与厨房以及书房与走廊之间的原有隔墙全部被打掉，使整个空间开阔舒适，浑然一体。厨房与书房均为开放式，客厅沙发背后透明的夹丝玻璃与厨房的中式屏风既划分了空间，又在整体上丰富了空间的层次。中式的设计元素在这里随处可见，客厅隔断上精美的祥云实木雕花，古韵十足的太师椅，典雅的中式落地灯，影视墙及餐厅墙面上造型独特的木雕，包括茶几上那别致有趣的茶壶与茶碗，无不流露着翩翩古风。设计在木上也古意盎然，高档的米黄云石地面，稳重的黑檀木，典雅秀丽的中式墙纸以及留白简化了的天花明确了空间的基调，而那偶然出现的一点红，一抹绿又丰富了空间的色彩关系。设计师用现代的材料营造着东方古韵，用现代的设计思路来处理每一样中式的元素与符号，用现代的设计语言来表达对一种文化的理解与感受。

　　步入主人房，绕过开放的洗手间，映入眼帘的背景墙以对称均衡的构图方式呈现在空间中。一片定做的真丝手绘花鸟墙纸仿佛在讲述着一个关于古时江南的优美故事，垂吊在两边的灯笼上散落着片片枫叶，红色的床头柜成为空间中耀眼的一笔。温馨的氛围，优雅的格调，让业主在忙碌了一天之后，躺在舒适的床上，拥着华丽的抱枕，做一个古色古香的梦。

Mu Xia

慕夏

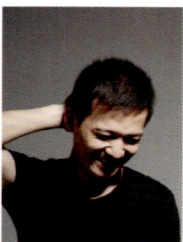

Design Company: Deep Design Consultants Co.,Ltd.

设计公司：厦门宽品设计顾问有限公司

Designer: Leo

设计师：李 泷

Location: Changsha

Area: 120m²

Main Material: grey wenge, grey wood grain marble, mirror, wallpaper

项目地点：长沙

建筑面积：**120m²**

主要材料：灰色铁刀木、灰木纹大理石、镜面、壁纸

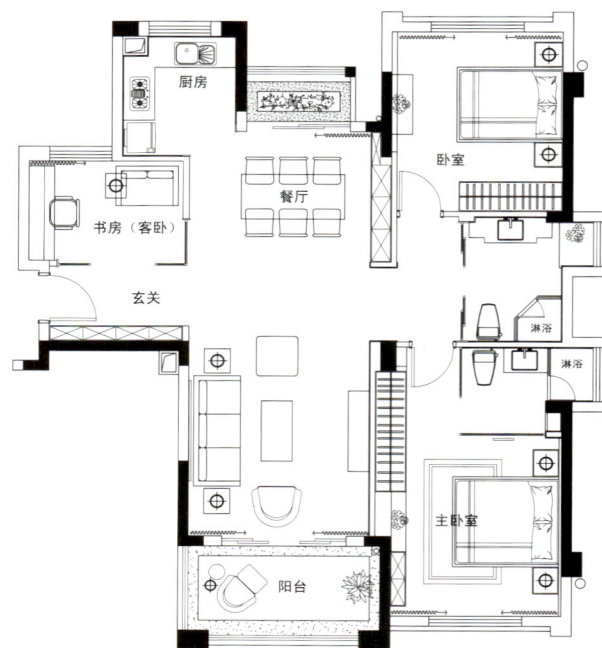

厨房

卧室

书房（客卧）

餐厅

玄关

淋浴

淋浴

主卧室

阳台

平面图

If the place you are back from work is full of fashion and artistic flavors, will you be more addicted to stylish residences? The project emphasizes the demand of fashion and strength. The designer used dark colors as the main tone, together with Gothic style metallic glitter, to create an ideal residence for urban dwellers. The metal along with crystal glass not only acts as enclosure but also displays the space. The first impression is the warm study which has white yarn curtain. Black, light grey and off-white constitutes the gradation of the room. The fine velvet which looks like the snow adds cool feeling to the space. Large area of warm wood color is accompanied with cool color materials, which makes owner has the feeling of wise and gentleness.

如果下班之后所回归的地方是一座充满了时尚艺术气息的生活美学馆，会不会使你对风格居所萌生更多的依恋？本案强调时尚与硬朗的风格诉求，设计师以深色为主调，搭配具有哥特气质的闪亮金属质感，营造都市新人类的理想空间。金属与通透清玻既围合空间，同时也展示空间。白纱飘逸的温暖书房呈现出空间的第一表情。深黑、浅灰、米白在空间中层叠出丰富层次，细腻的天鹅绒更是有夏日飘雪的意境，犹如迎面吹来的徐徐凉风。大块面的暖木色及素色布艺等具有柔软温度的软装搭与冷色系的冰冷感觉相互协调，让居者同时拥有了睿智出与柔和的感触。

Happiness Coast

幸福海岸

Design Company: Shenzhen Yipai Interior Design Co., Ltd

设计公司：幸福海岸

Designer: Duan Wenjuan

设计师：段文娟

Location: Shenzhen
Area: 200m²
Main Material: mosaic, solid wood, cloth art, plastic flower, accessories

项目地点：深圳
建筑面积：200m²
主要材料：马赛克、实木、布艺、仿真花、饰品

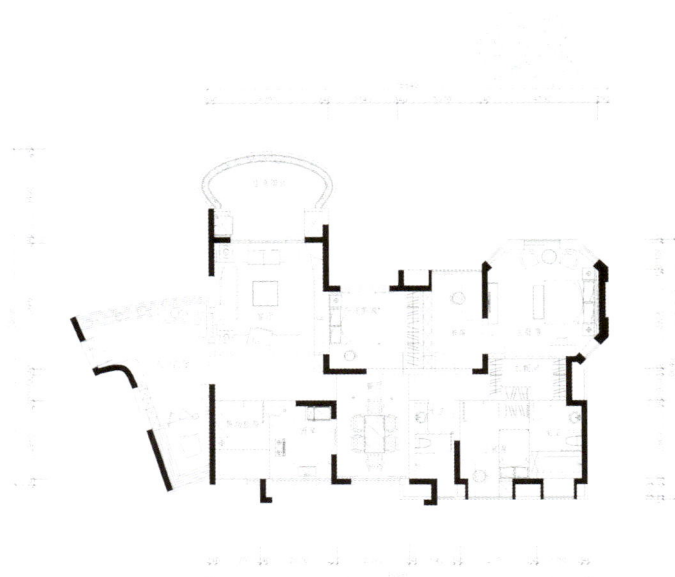

平面图

Originally the washstand and the bathroom were united. A wall was used to separate them. Moreover, a mosaic-decorated low wall was built between the stand and the aisle. In this way, the washstand outside and the toilet & shower facilities inside function separately, so that to avoid awkwardness. The color of the dinning furniture is the same pure white like the furniture in the living room. The waved ceiling is full of stereoscopic effect. The color of the chandelier coordinates with the patterns of the dinning furniture. The regular diamonds on the floor avoid monotonous feeling. The wine glasses and the candlestick are carefully-selected to not only match the design, but also display the bright feeling.

The master bedroom is in octagon shape. Almost each angle has a window. You can feel the sunshine flowing when you enter the room. The excellent lighting makes the colors of the subjects in the room look so vivid. The patterns of the wallpaper coordinate with the bed design—natural, elegant and low-profile refinement. The wood floor used on the bedroom distinguishes the one in the living room, and the wood grain conveys simple and warm feeling.

　　原本洗手台与淋浴间是一体的，设计师在中间用墙壁隔断，使洗手台单独成为一个功能分区，并且在洗手台与过道之间建了一道矮墙，表面饰以马赛克。这样的话，外间的洗手台与里间马桶和淋浴设施互不干扰，避免了尴尬。餐桌椅选用的是跟客厅家具一致的纯白色，餐厅天花交叉的波浪形造型极富立体感。吊灯色彩与餐椅上的花纹构成组合。地板上点缀着规则的小方块，使纯色地面不至于单调。餐桌上的酒杯和烛台无不经过了精心挑选，不仅要搭配，而且要表现出明亮的感觉来。

　　主卧呈较规则的八角形，几乎每一个角边都有一扇窗户，走进卧室便能感觉到阳光在流淌，良好的光线使室内所有物体的色彩都显得饱和。墙纸图案与床品很协调，自然、淡雅、含蓄、细致。卧室采用木质地板，与客厅加以区别，木纹本色感强，传达出简洁和温馨的感觉。

Decadentla

颓废寓言

Design Company: Jinzhan Interior Design Co., Ltd

设计公司：金湛室内设计有限公司

Designer: Ling Zhimo

设计师：凌志谟

Location: Taipei, China
Area: 210m²
Main Material: metal materials, glass, black steel

项目地点：中国台北
建筑面积：210m²
主要材料：金属材料、玻璃、黑钢

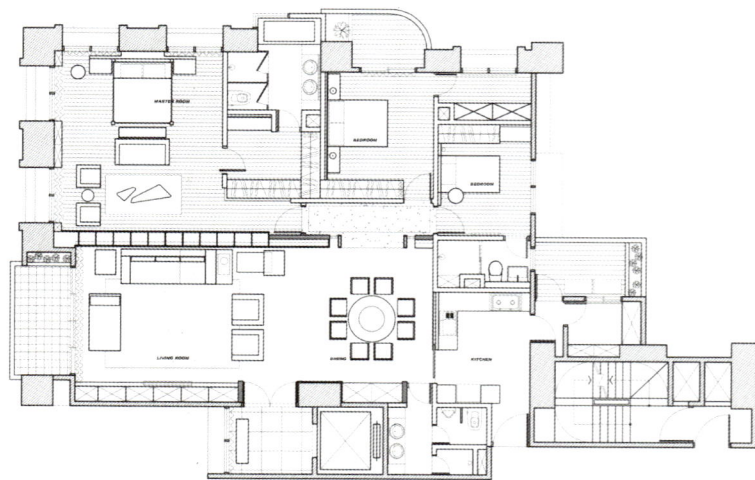

平面图

The philosophical concepts of Shostakovich, the famous composer in classical music history and the Latin word "Decadentla" are applied to the aura design of the space. Upon entering, one instantly feels the dark and decadent aesthetic of classical music. By blending decadent aesthetic with the proprietor's request for grand, chic style, a subtle but hard to conceal lavishness is born, flickering with elegance. The concept development of the space lies in a perfect presentation of the entire rectangular space, thus circulation flow is designed to vanish into the space, achieving a simple and whole sensory effect. unit sections, creating pure visual pleasure under the guidance of simple, instinctive linear vision. On the other hand, the designer applies the technique of contrast in color and shape: the mysteriousness and subtleness of antique black sets off the elegance of marble white; the mirrored walls reflect light, enclosing a grand, luxurious space; postmodern classical music pieces adorn the subtle and mysterious black carpet, integrating with the overall simple style of the living room, expressing individual style with perfection; the cool tone of the white television shelf, the warm tone of the stone and mirrored walls, and the black mirror titanium blend to eliminate overemphasis in a colorless space, attaining a balanced, tranquil effect. The irregular patterns on the black mirror titanium boards add rhythm, and in pursuit of aesthetic appeal over functionality demands. The designer plays with light and shadow by leaving a crack between the glossy black mirror titanium boards and the rough stone-painted walls behind the boards. Other than basic storage functions, the designer adds a special touch to the style design for the television wall of the master bedroom. Block-shaped waves increase variations in spatial layers, giving off an illusion of "shelf, not a shelf." The color of the carpet in the bedroom is, like the brighter-toned color of the floor, brighter-toned bluish lavender purple, producing horizontal and vertical visual planes which create a contrast between bright and dark in the overall space. In terms of space, mirror reflections create an illusion of a larger space, and together with the effect of light sources, the concept of point, line, and plane is fully employed in interleaved, reflected, overlapped and other lighting techniques, creating spatial aestheticism and demonstrating the uniqueness of the space.

设计将拉丁文 "Decadenta" 的哲学基础概念，转换运用在空间氛围设计表现上，在进入空间后便可感受到著名作曲家肖斯塔科维奇古典乐中神秘的美感。凇业主要求的华丽时尚风格糅合颓废美学，衍伸出低调却又难以掩饰的奢华，隐现出高雅的格调。空间概念的发展在于如何使整个矩形空间完美的呈现，因此在动线的设计上使动线消弥于空间，达到简约而完整的惑官尺度。华丽古典的对称手法在餐厅与私密领域入口完整呈现，左右两侧一连串黑金属质感的水平横带造型，使空间不被单元机能所分割，在简单且直觉的线性视觉引导下，营造出纯净的视觉享受。对比手法被巧妙地运用在颜色与造形上，神秘又带低调的古典黑衬出雕刻白大理石的高雅，墙壁镜面质感反射光线亮点，围塑出大方奢华的空间感。后现代古典乐章铺陈在低调又带神秘的黑地毯上，与客厅整体的简约风格融合，完美的演绎出自我的格调。冷色系的白色电视柜、暖色系的洞石镜面墙与黑镜镀钛金属的调和，使无彩度的空间不会过分地被强调并达到安定平衡的效果。黑镜镀钛层板的不规律变化，使空间更有节奏感，且追求美感的程度大于机能需求。因此，层板后排糙石头漆材质的墙面与光滑黑镜镀钛层板间留有缝隙，丰富了光影的变化。主卧室的电视墙面除了基本的收纳机能外，在造形上也有特点。块状的起伏造型使空间的层次更富有变化性，不禁令人产生有"柜，非柜"的错觉。卧室地毯选用和地板一样明度较高的薰衣草色系，在整体空间的视觉上产生水平面与垂直面明与暗的对比。空间上除了大量运用镜面反射的原理创造出更大空间感的错觉外，加上光源效果的处理，充分地运用点、线、面原理，展现出光在交错、反射、重叠，创造出空间的唯美性与完整显现空间的独特性。

Ju Yue

掬月

Design Company: Deep Design Consultants Co,.Ltd.

设计公司：厦门宽品设计顾问有限公司

Designer: Leo

设计师：李泷

Location: Changsha
Area: 130m²
Main Material: violets, moonlight marble, mirror, wallpaper

项目地点：长沙
建筑面积：130m²
主要材料：紫罗兰、月光大理石、镜面、壁纸

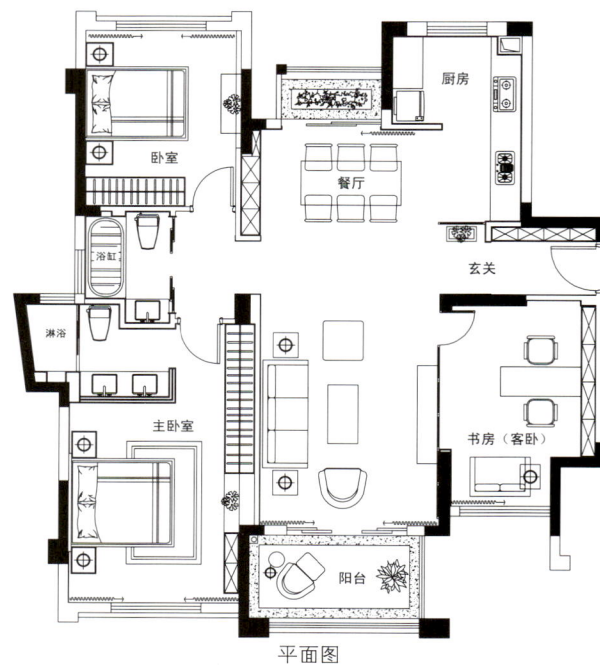

平面图

After constantly pondering and simplifying, the composing of poem reflects the heart and the apperance, transcending the framework of the words. So in the space being designed, the existence of each element is placed in a proper position after continous deliberation and discretion, It is the personal thought of the designer after refined, condensed, integrated and organized. The derived emotion transcends the design itself.

As the case extracted from the tang poetry but fusion with lifestyle. Autumn is the season of mature. The stable tone of this case echoes with autumm at a distance, comfortable and meaningful. The semi-permeable and semi-transparent golden glass partition is like the autumn leaves, revealing the implicit and elgant eastern beauty gleamingly. The warm wood color and the soft touch of the exquisite cloth art create the delicate and soft life, composing the life scenery which people taste with appreciation.

　　如果说一首诗里字与字之间的构组，在不断地琢磨和精简之后，反射出心灵图像与外观画面往往超越了文字之间的框架。那么在经过设计的空间里，每个元素的存在皆通过不断地斟酌与推敲之后，才被放置在一个恰到好处的位置，都是设计者个人思维经过提炼、浓缩、整合，组织之后，衍生出的情感，超越了设计本身。

　　如同此案提炼自唐诗却融合了生活方式。秋季是成熟的季节，本案沉稳的色调与之遥相呼应，安逸而隽永。半隔半透的金色夹纱玻璃宛如秋天飘落的枯黄树叶，隐约透着含蓄优雅的东方美感。温暖的木色与柔和触感的精致布品营造出精致柔美的生活温度，构成引人吟味的生活风景。

Realm of the Quiet

静谧之境

Design Company: Jinzhan Interior Design Co., Ltd
设计公司：金湛室内设计有限公司
Designer: Ling Zhimo
设计师：凌志谟

Location: Taiwan, China
Area: 150m²
Main Material: air dry wood, paint, artificial stone-paint, glass, marble, accordion-shaped curtain, oak

项目地点：中国台湾
建筑面积：150m²
主要材料：风化木、喷漆、人造石、仿石漆、玻璃、大理石、风琴帘、橡木

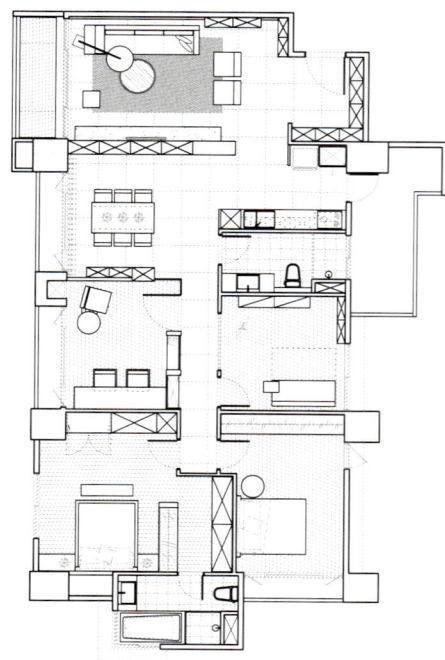

平面图

Combination of the virtual and the real is hidden in space. There are no fancy media nor luxurious decorations. The beauty of nature and quiet is derived from the pureness of space. For public area, the living room and the dining room are divided merely by the segmentation effect of the ceiling, which offers not only the distinction of designated areas but also the space continuity at the same time. Indirect lighting on the ceiling connects the internal space (private rooms) with the external one (public area). More than just a simple conduit, the aisle creates a sense of visual penetration with the combined design of the virtual and the real on both walls along the aisle. The TV wall is designed with asymmetrical balance, presenting the beauty of irregular layers. Two sides of the installation suggest the possibility of a space dialogue. Functional designs on the front and back sides create a sense of interaction between furniture and space. Three originally independent units of space, the studio, the study and the dining room, are arranged with the backdrop of glasses and mirrors. While disguising the storage spaces, the installation of glasses and mirrors also creates a neat and stylish design as well as an interrelated sense of visual penetration. The alternate use of stone bricks and wood softens the contour of the interior space, exuding the warmth of humanity and adding the variation of layers to overall design.

　　虚与实的交错，隐藏在空间里。没有炫丽的媒材，没有奢华的装饰，而是将空间的纯粹衍生出自然静谧之美。打开公共空间，客餐厅空间仅利用天花板区域分割的效果，使空间不仅独立却也链起空间的串连性。天花板的间接照明，连贯着由内（私密空间）到外（公共空间），其中的走道动线不再只是走道，两侧的墙面虚与实的分割设计，创造出空间接口上的视觉穿透性。电视矮墙非对称性的平衡表现出不规则的层次之美。体量的两面，发展出空间对话的可能性，前后两侧的机能设计，让家具与空间有了互动。工作室、书房、餐厅，三个原本独立的单元空间，应用玻璃与镜面的穿插，衬托出利落简约感，也隐藏了收纳机能，穿透效果也因此有了彼此的背景。砖石质感与木质的交替作用调和了室内线条，呈现出人性的温度，更为整体空间增加了层次的变化。

Tangquan Golf A1 Sample Room

汤泉高尔夫A1样板房

Design Company: Shenzhen Persons Environmental Art Design Co., Ltd

设计公司：深圳市墨客环境艺术设计有限公司

Designer: Wang Qinjian

设计师：王勤俭

Location: Huizhou
Area: 265m²
Main Material: oak, carpet, marble, leather, matt golden foil, wallpaper

项目地点：**惠州**
建筑面积：**265m²**
主要材料：橡木、地毯、大理石、真皮硬包、亚金箔、墙纸

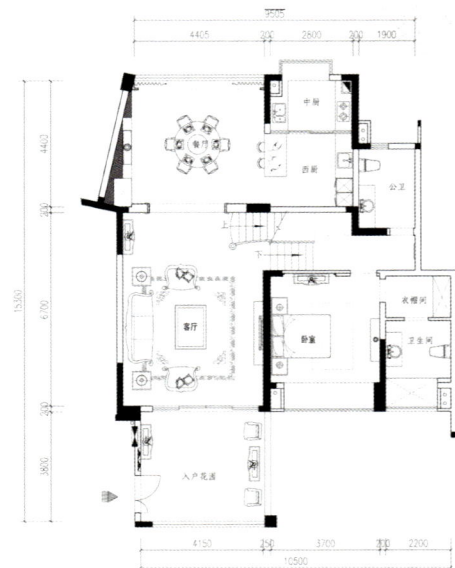

一层平面图

Design inspiration comes from the new age Renaissance, Combining modern and Classical elements. Using white as the main tone, with the soft lighting and the gentle texture, the meticulous, gorgeous white marble and the exquisite champagne foil are the basic elements. Each detail is superior to show the luxurious space, cultured architectural elements increase dimensional aesthetics, exquisite and refined decoration sends out a unique art accomplishment.

Precious thoughts are revealed everywhere. The noble taste and the luxury life achieve a balance. The case shows the magnificent and in the meantime shows the elegant feeling.

设计灵感来自新时代的文艺复兴，结合了现代与古典的元素。以白色为主色调，在柔和光线及柔滑质感的结合下，细致华丽的纯白大理石、细腻的香槟银箔是基本元素。每一细节都展示出空间的奢华，讲究的建筑元素增加了空间的美感，细腻雅致的摆设散发出一份别具慧眼的艺术修养。

处处显露心思，弥足珍贵。高尚格调的品位与奢华的生活达至平衡点，华贵由此璀璨盛放，同时了展现了优雅的一面！

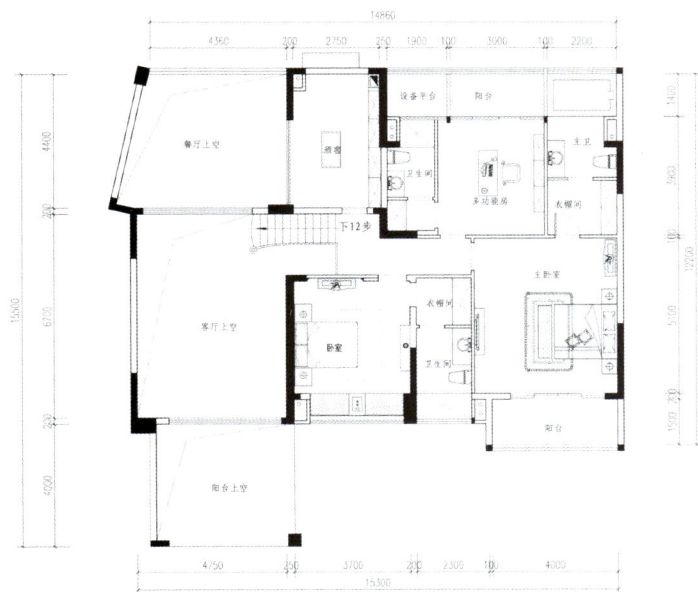

二层平面图

The Fifth Park, Vanke

万科第五园

Design Company: Shenzhen Wang Wuping Design Studio

设计公司：深圳五平设计机构

Designer: Wang Wuping

设计师：王五平

Location: Shenzhen
Area: 230m²
Main Material: Dulux paint, polishing brick, grey mirror, metal mosaic

项目地点：深圳
建筑面积：230m²
主要材料：多乐士乳胶漆、抛光砖、灰镜、金属马赛克

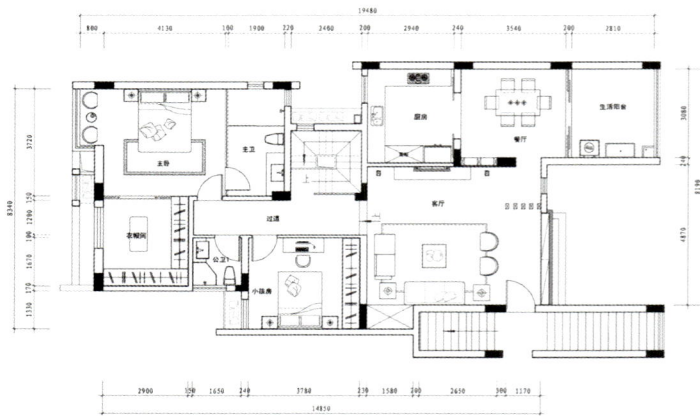

一层平面图

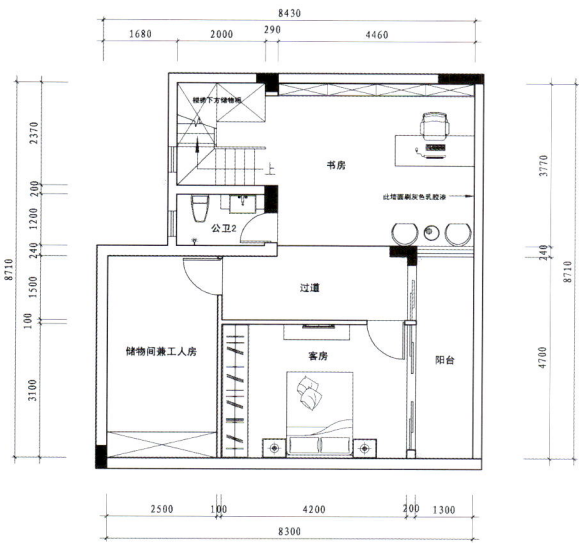

二层平面图

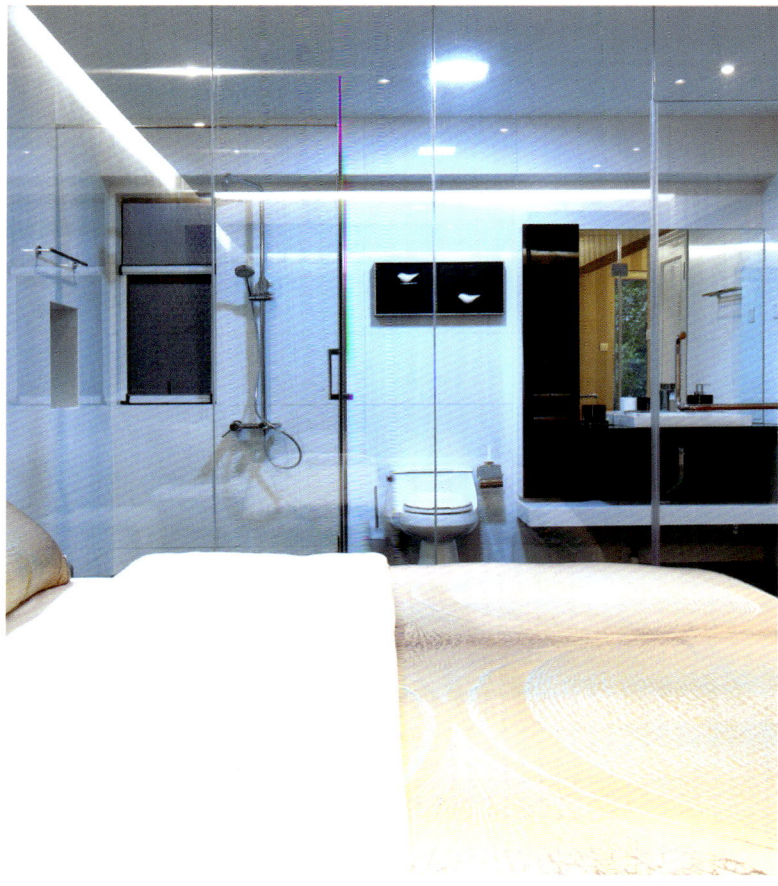

New Shoreline

新岸线

Design Company: Shenzhen Yipai Interior Design Co., Ltd

设计公司：深圳市伊派室内设计有限公司

Designer: Duan Wenjuan

设计师：段文娟

Location: Shenzhen
Area: 145m²
Main Material: solid wood, cloth art, artificial flower, plant, rustic tile, mosaic

项目地点：深圳
建筑面积：145m²
主要材料：实木、布艺、仿真花、植物、仿古砖、马赛克

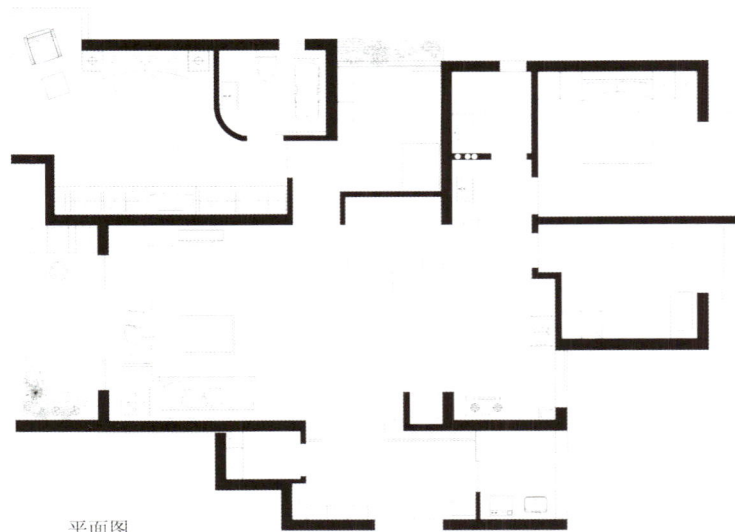

平面图

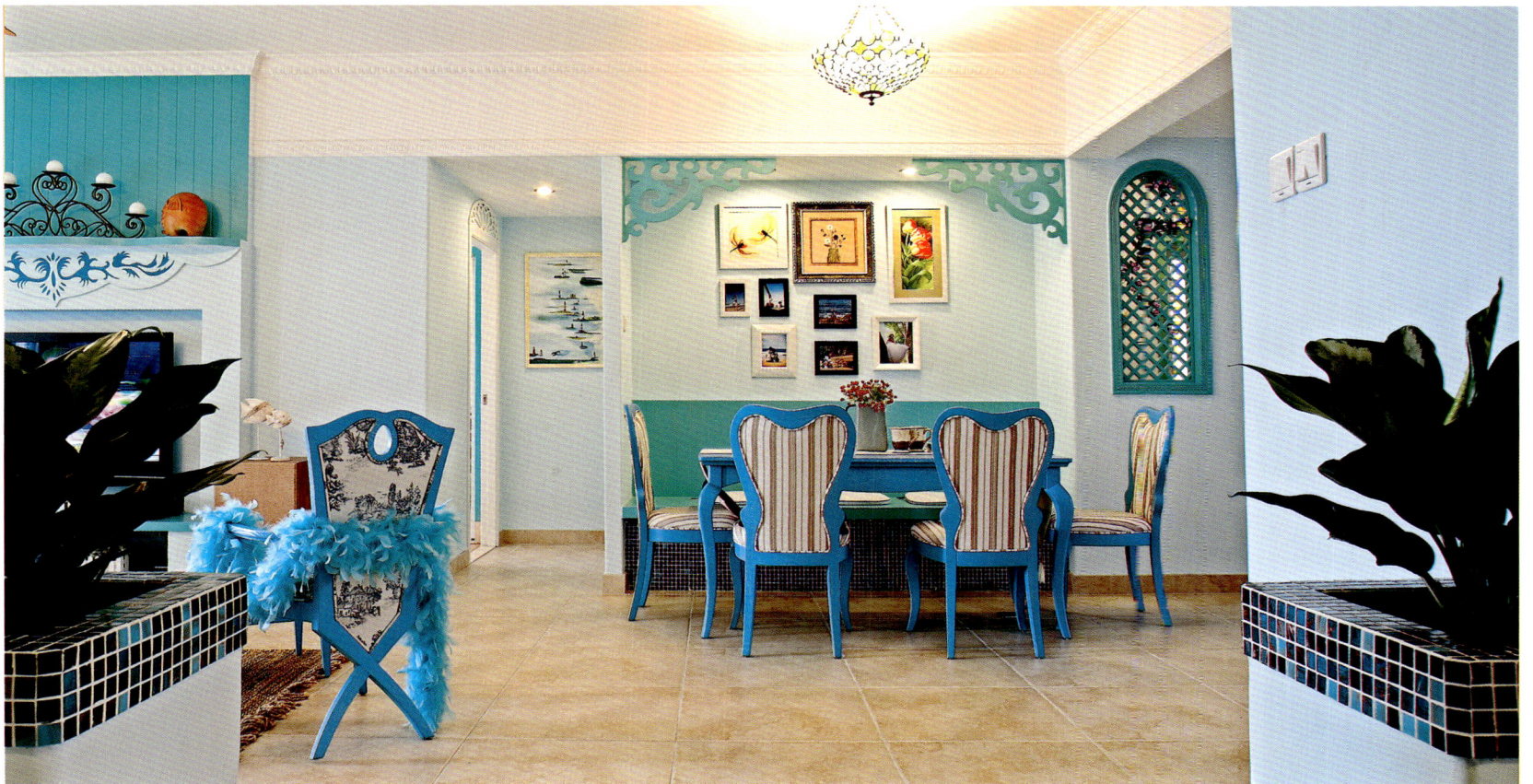

Mix & Match is always referred to these years in the fashion field. The fashionable youngsters are especially interested in trying new thing and bardian styles. Hence, this house is designed for the elegant fashionable people who love life. These people are fond of traveling, shooting, music, etc. They are inclined to simple and free life style.

Beside the fresh plants in the designed flower bed, there is a lovely decoration with the warm word "welcome" in it, on which the host could hang the keys and start to prepare a wonderful mixed meal with Chinese and Western elements. The figure made up of mosaic in the carpet not only well decorates the space but also separates each function area. Feather enwound around the sofa makes a warm atmosphere. Entering the room, you can feel the fresh lively stream of life.

The water globule mosaic is fabulous on the wall of the main bedroom. To a living space, it is all to add into some specific visual attraction rather than large scale of design in the existing environment on condition that functions have been reached.

混搭的概念近年来经常被时尚界所提及。特别是对于一直走在潮流前沿的年轻人来说，他们更喜欢尝试新事物，接受个性风格。所以这套户型作品针对的正是喜欢娴静、优雅，懂得生活的时尚达人。他们热爱旅行、摄影、音乐，喜欢简单、自由自在的生活……

　　入户区设计的花池里生长着生气盎然的绿植，旁边十分具有童趣的装饰品写着温馨的"welcome"（"欢迎回家"）。主人随手将钥匙挂在上面，便可以进入开放式厨房，准备一顿中西混搭的美餐。地面上用马赛克拼出的地毯图案既有装饰性，又有分区的功能……轻轻缠绕在客厅椅上的羽毛让人感到生活的温暖。进入这个空间，你马上就能感受到浓浓的生活气息在流淌。

　　主卧入口墙壁上的水珠马赛克令人惊艳。对于一个居住空间来说，在已有环境的基础上，满足功能之余倒并不需要大刀阔斧地设计造型，仅仅需要一些细节上的视觉亮点，已经足以打动人心。

Warm Jade

暖玉

Design Company: D-Hand Design Firm(Fuzhou)

设计公司：鼎汉唐（福州）设计机构

Designer: Chen Mingchen

设计师：陈明晨

Location: Fuzhou
Area: 140m^2
Main Material: monto paint, laminated floor, rustic tile

项目地点：福州
建筑面积：**140m^2**
主要材料：蒙托漆、金刚板、仿古砖

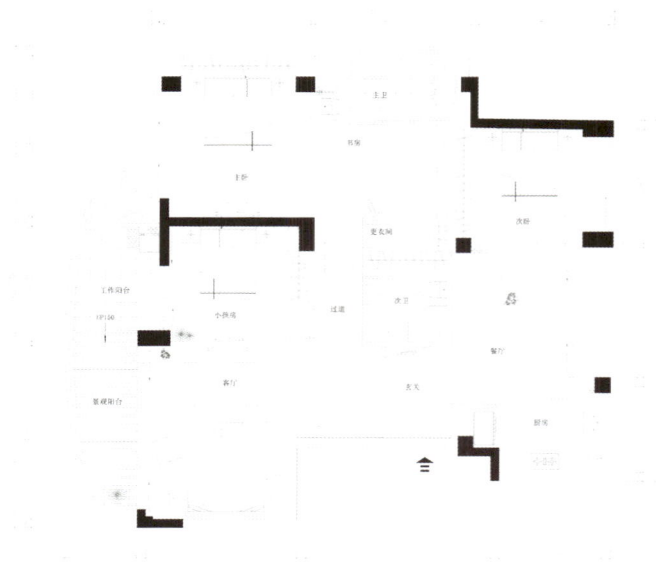

平面图

Volume

空体

Design Company: Goldesign Studio

设计公司：金湛室内设计有限公司

Designer: Ling Zhimo

设计师：凌志谟

Location: Taiwan, China
Area: 250m²
Main Material: oak, weathered wood, maple, teak, stone paint, glass, travertine, stainless steel, leather, fabric

项目地点：中国台北
建筑面积：250m²
主要材料：橡木、风化木、枫木、柚木、仿石漆、玻璃、米洞石、不锈钢、皮革、布料

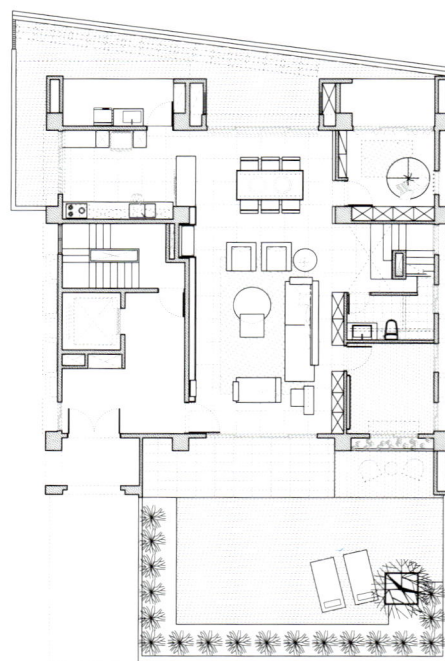

平面图

In accordance with the "Architecture as Volume" principle promoted by the school of the International Style, the design is created to maintain the complete volume as well as the spatial mobility, representing the Modernist architectural aesthetics of the 1930s. The fragmented spaces in the volumes are integrated. Volume is not sharply divided by functional designs and all unit spaces are overlapped accordingly. A sense of natural and leisure life is introduced into the Modernist International Style of the volume. The simple lifestyle is to resonate with the leisure lifestyle in one space, and a more pleasant living space is thus evolved. The pursuit of simplicity and nature in life can be achieved.

Completeness of a large space is retained with the overlapping effect created by the marble TV wall and the back wall decorated in large blocks of volume and color in the living room. The design makes a double-layer effect and freer lines of movement possible in the space. Rectangular volumes are converted in the horizontal and vertical spaces. The vertical line of movement is transformed into volume, telling the stories of space with light and shadow. The low-key elegance is merged with the leisure style to create a variety of changes in spatial levels. The beams of the original building in the master bedroom are used to divide the spaces in accordance with the "no decoration" guideline.

依循国际风格"建筑像体量一样"的原则，设计试图保留完整的体量，并保持空间流动，呈现 20 世纪 30 年代的现代建筑空间美学。整合体量空间之间所产生的零碎的空间；不使体量被空间机能所划分，单元空间因而相互重叠。在纯粹体量的原则下，试图将自然休闲的生活感引入国际风格当中，而不失现代主义风格。简约风与休闲风的生活质感，在空间里产生共鸣，演化出更宜人的居家空间，追求生活中简约与自然的质朴淬炼。

客厅背景墙运用大面体量和色块的手法与电视墙的大理石墙两者前后互相重叠，因而保有大尺度空间的完整感，使空间产生双重层次的效果，让动线自由于空间中。矩形量体在水平与垂直的空间中转换，将垂直动线发展成体量表现，在光影幻化下讲述空间故事，融合低调素雅与休闲的风格，创造丰富的空间层次变化。主卧房运用建筑体的横梁构造，在无装饰的原则下，划分出不同的空间。

Jia He Sheng Shi

家和盛世

Design Company: Hwayon Interior Group
设计公司：深圳华空间机构·华旸商业设计
Designer: Xiong Huayang
设计师：熊华阳

Location: Shenzhen
Area: 200m²
Main Material: marble, polishing tile, mosaic, wallpaper, laminat floor, matt glass

项目地点：深圳
建筑面积：200m²
主要材料：大理石、抛光砖、马赛克、墙纸、复合木地板、磨砂玻璃

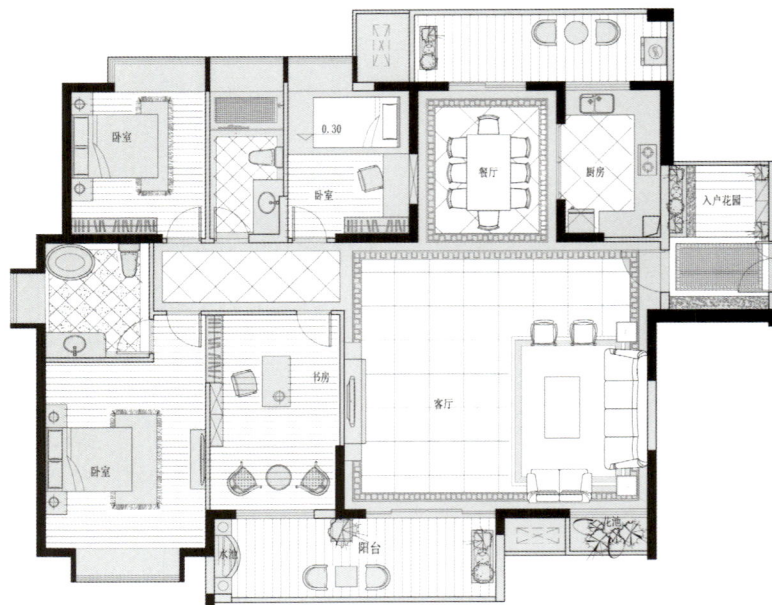

平面图

People-oriented now has become the new definition of house design. Neoclassicism is a kind of multi-level art of luxury and extravagance. Is luxurious and extravagant decoration suitable for residential space? After all, it is human being that lives in this house, it is human being that experiences the conveniences or inconveniences of life. Therefore, the designers abandoned the complex furniture, and used simple and modern techniques to build the neoclassic design for this project.

The target customer of "Jia He Sheng Shi" is the white-collar family in middle or high class. They are keen on leisure and comfortable family life when they get off work. Hence, in terms of function division and space design, the designers paid more attention to the openness of public space and the concealment of bedrooms. The living room has spacious room so that the family member can freely enjoy themselves.

The color of neoclassic style is steady with a little exaggeration. It is either in bright red, or deep black or brown. The project used black, brown and silver as the main color for the decoration. The wall was just slightly decorated and without any glaring excessive "make up". Therefore, it maintained the original pure and fresh feeling. The whole space was surrounded by elegant atmosphere.

The open dinning room is undoubtedly the best choice for family member's communication. A dinning room which is full of happiness next to the garden with greenery scenery make you enjoy the food and the life at the same time.

Private space such as the study and bedroom required quiet atmosphere. There is no exaggerated design or over-decoration. One less would be too less while one more would be too more. Every detail is in balance and perfect. When you lean on the sofa and look at your family's pictures in the afternoon sunlight, you will feel happy and contented. No matter how hard the work is and how much cost you have to pay, it deserves as long as you have a family which is your happy harbor.

以人为本，现已成为样板房设计的新定义。而新古典给人的感觉，总是奢华与繁琐的多层次艺术。但是奢华与繁琐的装饰是否适合家居生活呢？毕竟，是人们生活居住在这个家里，是人们在体验着生活的或繁琐或方便。所以设计师摒弃繁琐的家居，运用简约与时尚的手法构成本案的新古典主义设计。

"家和盛世"的目标客户是中高端的白领家庭，她们在工作之余，追求舒适自然的家庭生活。所以在室内功能划分及空间设计上，倾注于公共场所的开放性与卧室的隐密性。客厅需有宽阔的空间使家庭成员不受拘束，自由且随意。

新古典风格的色彩就是在深沉中略有夸张，或是浮华的红，或是幽沉的黑或咖啡。本案以黑色、咖啡色、银色三种相近色为主要装饰。墙壁只是略做修饰，并未有炫目壁纸的过分"化妆"，所以还保持着原本的清新淡雅。整个空间拥有高雅的氛围。

倾注于家庭成员的交流沟通，开放式餐厅无疑是最好的设计。一边毗邻幸福满溢的客厅，一边毗邻绿意盎然的小花园，让你品味食物之时也品味生活。

书房及卧室等个人空间，需营造静谧的气氛。没有夸张手法的设计，也没有过分的装饰，少一件则空，多一件则满，恰当好处才是最完美的。在午后的阳光下，斜靠在沙发上，翻起往日与家人游玩的照片，或许你不禁涌出一种幸福感。无论多么努力的工作，多么辛苦的付出，都是值得的。有了家，就是有了幸福的港湾。

Han Bi

涵碧

Design Company: Deep Design Consultants Co.,Ltd.
设计公司：厦门宽品设计顾问有限公司
Designer: Leo
设计师：李 泷

Location: Changsha
Area: 80m²
Main Material: white wood floor, white travertine, wallpaper, clear mirror

项目地点：长沙
建筑面积：80m²
主要材料：白影木地板、超白洞石、壁纸、清镜

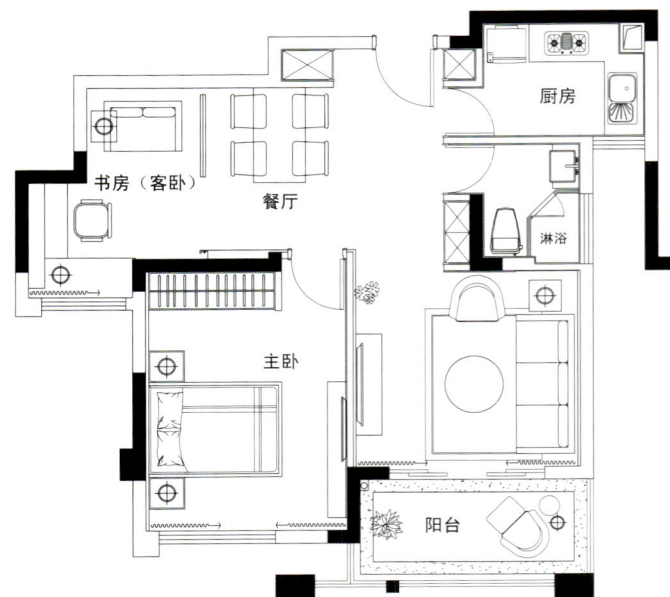

书房（客卧）
餐厅
厨房
淋浴
主卧
阳台

平面图

The key to small space interior design, not only in the originality of shaping space, but also should pay more attention to integral unity feeling and the coordination, making people feel the simple sense of the space and the clear essence of life.

This case is a small unit with one room, one dining room and one living room. The whole design uses beige as the main tone. The designer hopes to construct a sweet and pure habitation. The details of the space are simple, with no extra adornment, presenting a clean and mild atmosphere, filling with elegance. The design emphasizes contracted and softhearted design. The feeling of home is not created by fancy visual stimulation and decoration, but should focus on the real quiet and comfortable of people.

　　小空间的室内设计重点，不仅在塑造空间的创意上，更应该注重整体的统一感与协调性，让居者感受到空间的质感以及生活清澈的本质。

　　本案为一室两厅的小户型，整体采用轻盈温暖的米白色调，设计师希望借此构建一个温馨纯净的心灵居所。空间的细部精简，没有多余的装饰，内部的气氛洁净而温和，充盈着优雅气息。设计师强调简约而温情的设计，家的感觉不在于花俏的视觉刺激及装潢，而应着重让居者可以享受真正的宁静与舒适。

Triumph Garden

凯旋花园

Design Company: Kern & Roy Urban Planning and Landscape Design Ltd.

设计公司：柯恩罗伊景观设计事务所构

Designer: Teng Jiafeng

设计师：滕家锋

Area: 330m²
Main Material: oak, marble, glass

建筑面积：330m²
主要材料：橡木、大理石、玻璃

一层平面图

Illustrating the Connection Between Life and Space

Here, you can examine the rationality of space and its connection with life style from an objective point of view. The impersonal angle refers to the house owner's family; life style refers to the living style that the family members expect. The designer illustrates the connection between life and space through his design.

Reinforcing Plane

When it comes to design, people also need to review the existed floor plan. The designer needs to build the connection between the space nature and living habits in order to search for the relationship in terms of space's movement & stillness, solidness & voidness, open & enclosure. The owner's life style is connected with the planning. Eventually, the main body and object are formed and then reinforcing relationship will be derived from the design.

The Continuity of the Elevation

In addition, the connection with space has to be achieved via vertical surface. Movement & stillness, solidness & voidness, open & enclosure constitute to vivid and clear pictures. The overlay of vertical surface creates a kind of visual shading, which makes a united and floating space.

The Use of Symbols

Symbols and graphics can't be ignored in the whole space. They are infintely magnified and emphasized. You will be convinced that it is not simply the mixture of symbols. It is the illustration and interpretation for a kind of life style.

诠释空间与生活的关联性

在此，你可以尽可能地从客观的角度上检视空间的合理以及与生活形态的关联。这里所说的客观角度是指住宅主体的家庭成员结构，生活形态则是指住宅主体所适应的或是希望的生活方式。设计师便是借由空间设计来诠释这二者的联系。

递进式的平面

设计时，还需要重新审视固有的平面配置，构建单元空间性质与生活习惯连接，并利用平面配置来连接业主的生活形态，以寻找空间的动静、实虚、开合之间的对应关系。如此这般，也就形成了空间中的"主体"与"客体"，并衍生递进式的层次关系。

立面的连续性

此外，设计还借由立面的形态来完成单位空间的连接，通过动静、实虚、开合充分展开，形成一个动态的清晰画面。并通过立面及叠加完成视觉上的遮挡性，塑造出一个整体而流动的空间。

符号的使用

在整个空间中，式样符号不容忽视。它们被无限放大，并加以强调。你必须相信，这绝非简单的符号穿插，而是一种生活形态的标榜和诠释。

Dongguan Dream • Qinyuan Residence

东莞理想•沁园

Design Company: Shenzhen Wang Wuping Design Studio

设计公司：深圳五平设计机构

Designer: Wang Wuping

设计师：王五平

Location: Dongguan
Area: 260m²
Main Material: Dulux paint, polishing brick, wallpaper, crystal lights

项目地点：东莞
建筑面积：260m²
主要材料：多乐士乳胶漆、抛光砖、墙纸、水晶灯

一层平面图

Rational space planning, fine combination of inflexibility and yielding, simple and brief design techniques, pure color and fashion complement each other. With the fine design furniture, each detail reveals the purchase for the perfection of life.

The special part of this project is the top part of the ladder space which is used as part of the kitchen where the fridge hides in. It greatly reduces the pressure of the originally small kitchen. The owner wanted a subtly luxuriant style. Therefore, when it came to design, the designer made it simple and modern; when it came to decoration such as lights, furniture, accessories, the designer chose the fashionable and noble ones.

本案不折不扣的空间规划，刚柔并济的形体混搭，简洁流畅的设计手法，纯色与时尚相辅相成，并配以设计感十足的家具，无不透视出对生活完美极致的追求。

在平面规划上，本案的巧妙之处就是把楼梯空间最高位部分给了厨房，正好把冰箱藏了进去，这样就大大减轻了本来就不大的厨房的空间压力。由于屋主要求的设计风格为隐约的华丽，所以在格调定义上，尽可能做到简洁现代的硬装，而在灯光、家具、配饰上要做到时尚，贵气。

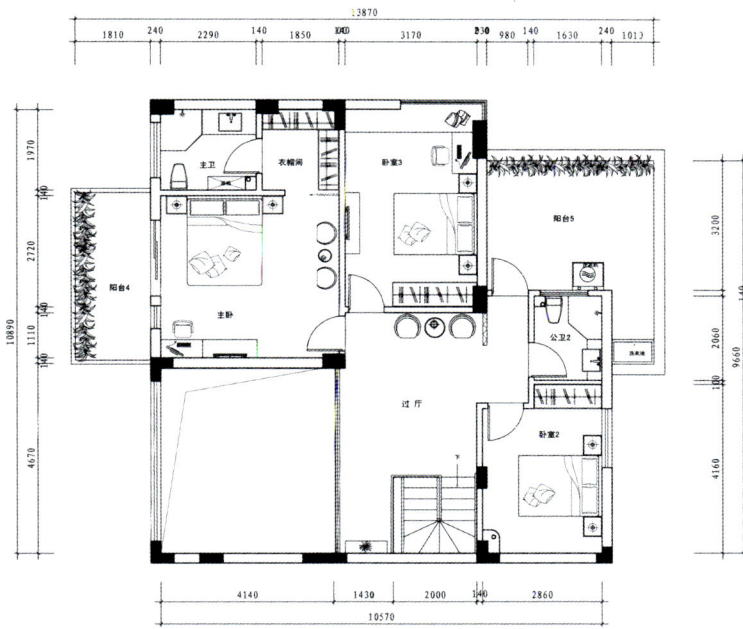

二层平面图

Portofino • Water Region

波托菲诺•纯水岸

Design Company: Shenzhen Wang Wuping Design Studio

设计公司：深圳五平设计机构

Designer: Wang Wuping

设计师：王五平

Location: Shenzhen
Area: 600m²
Main Material: grey mirror, red oak, wallpaper, polishing brick, light emperador marble, grey wood grain marble, mosaic

项目地点：深圳
建筑面积：600m²
主要材料：灰镜、红橡木、墙纸、抛光砖、浅啡网大理石、灰木纹大理石、马赛克

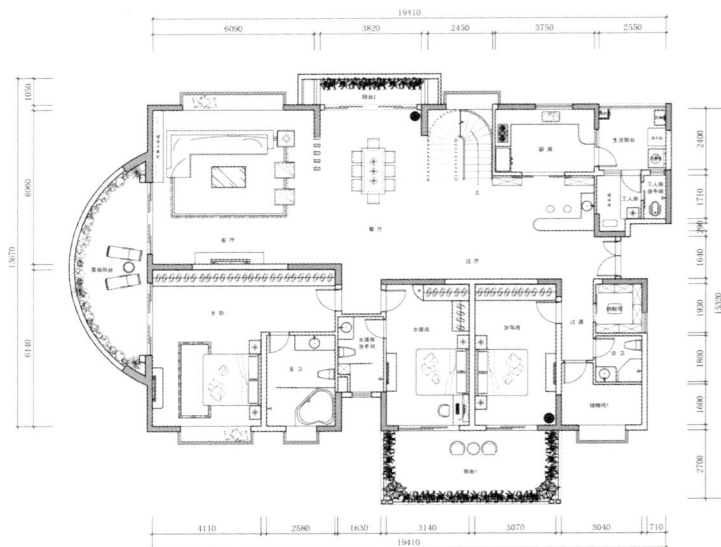

一层平面图

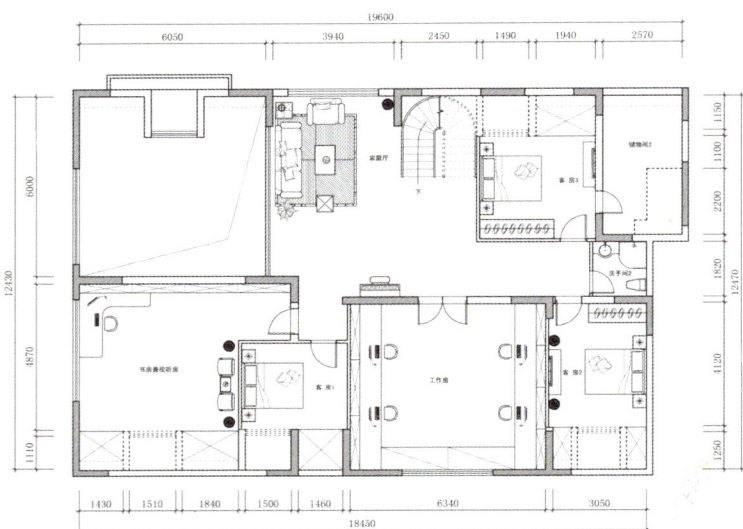

二层平面图

The owner is a sunshine and chipper man. He wanted his house to be the same kind, and could be achieved through design.

It is a top storey compound house with complicated and diversified appearance. The second floor has ample space, so the stairway requires to be reset. Only reasonable design and allocation of the space will ensure that there is no dark room or low functional space. The owner had been worrying about that. Understanding what the owner wanted, the designer worked out this project for him.

The diversified ceiling led to a flexible design technique. Part of the geometric shapes was kept, and some special design ways were added to make it fit perfectly with the major part of the house. The decoration aims to create a special space which has the feeling of softness with hardness hiding in.

屋主性格阳光爽朗，希望通过设计让他的新家也带有这样的属性。

由于是顶层大复式结构，洋房的外观及顶面复杂多变及斜异相间。二楼是一个大开空间，楼梯空间需要重新定位。只有将空间设计分配合理化，才不会存在暗房和低点功能空间。这些原始的结构形态，一直困扰着屋主。设计师结合了屋主的一些想法，站在他的立场，融入自己的专业，生成了这个方案。

异形多变的天花形体，决定了设计手法要依形就势，不必刻意的去处理，保持部分的几何天花，并在立面形体上也采用一些建筑结构语言的设计手法，与之呼应，与本案大格局的空间相得益彰。在装饰效果上，立求一种刚柔并济的感觉，营造一个气宇不凡的空间。

Yasong Residence

雅颂居

Design Company: Shenzhen Wang Wuping Design Studio

设计公司：深圳五平设计机构

Designer: Wang Wuping

设计师：王五平

Location: Shenzhen
Area: 300m²
Main Material: art glass, red oak, wallpaper, imitational marble brick, grey wood grain marble, mosaic, crystal lamp

项目地点：深圳
建筑面积：300m²
主要材料：艺术玻璃、红橡木、墙纸、仿大理石砖、灰木纹大理石、马赛克、水晶灯

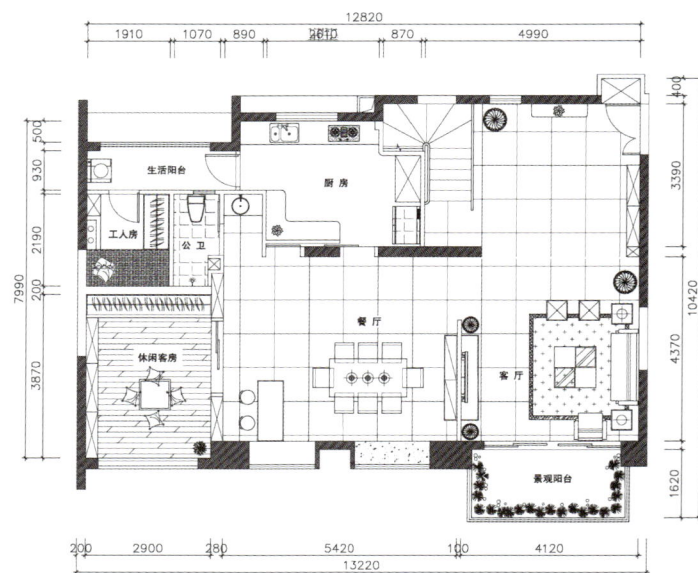

一层平面图

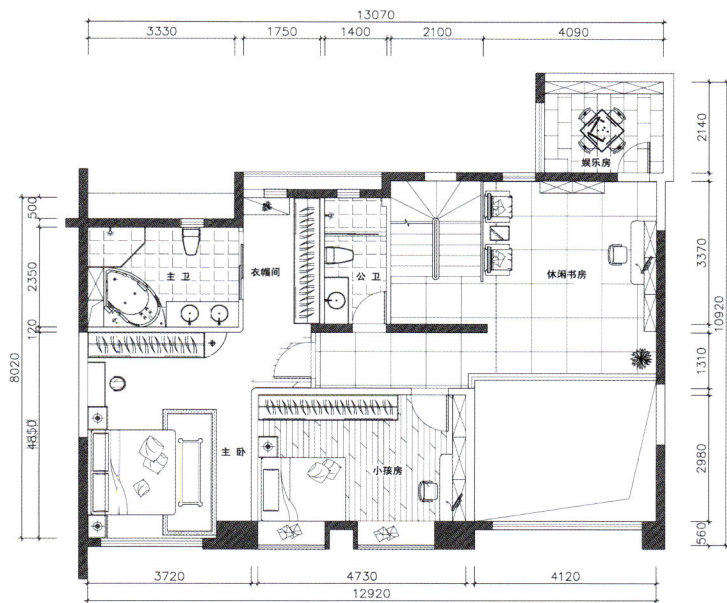

二层平面图

Every detail in this house indicates the owner's romantic and aesthetic purchase towards life.

The owner wanted her house simple and bright, and has an elegant and delicate feeling.

Once the theme was decided, everything ran smoothly. The design is combined with modern and European elements. The main color is white. Some furniture and accessories are in dark color, which enrich the space. The straight line in the background wall of the living room makes the space look more ample.

这里的每一个细节都畅想着屋主对生活的浪漫唯美的追求。

接手委托设计时，屋主说，要让家变得干净明快，有点飘逸淡雅的感觉。有了主题，一切都不再困惑了，接下来便开始了马拉松式的细节推敲选材工作。

本案运月混搭的设计手法，融入了现代、欧式的设计元素，采用白色为主色调，局部配以深色的家具和软装，让空间层次丰富起来。客厅背景直线条的运用，使得空间显得更加大气。

Landsea International Block
朗诗国际街区

Design Company: WILLIS Design Studio

设计公司：巫小伟设计事务所

Designer: Wu Xiaowei

设计师：巫小伟

Location: Suzhou
Area: 130 m²
Main Material: Hong Kong MJ furniture, solid wood flooring, Kohler Kitchen, Kohler sanitary ware, solid wood shutter doors, wallpaper

项目地点：苏州
建筑面积：130m²
主要材料：香港美兆家具、实木复合地板、科勒厨房、科勒洁具、实木百叶门、壁纸

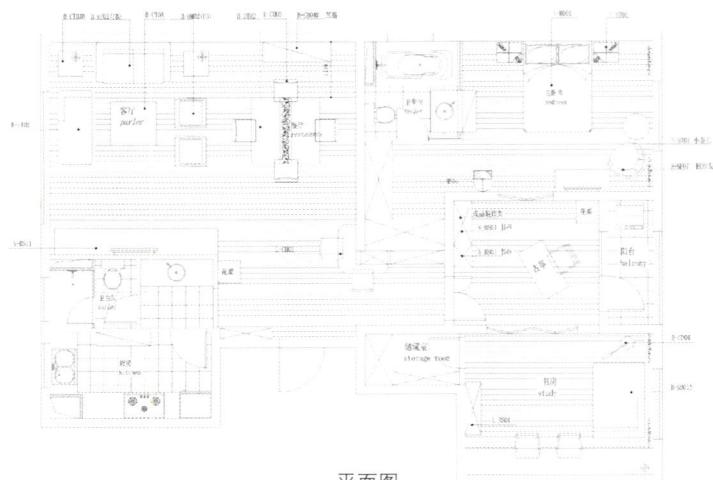

平面图

This case is a flat of Landsea International.The designer further clarifies the function of different areas on the basis of the orignal structure, and makes it more reasonable with slightly modified. The case locates in contemporary and simple style, using black and grey as fundamental key, through the depth change of black, white and grey to reveal the art and charm of mordern life. The ceiling of the house is barely decorated, simply using plaster and downlights, concise and without losing the sense of layers. The wall also is just decorated with modern styled decorative painting, filling the visual effect of the space. The sitting room has a fabric sofa full of sence of lines; part of the TV background wall sets the lockers and cabinets, and shelves are placed on the wall, guaranteeing the storage function of the living room. The reception area and the study area adopt pannel furniture. The fresh and bright style brings unique feeling to the space even with no more decorating. The designer designs a small bar in the transition area between the living room and the study room, one can enjoy tea or wine as his/her own wish. The wall of master bedroom uses the purple decorative wallpaper, and the some colored curtains. Purple brings noble, warm, mysterious and classic atmosphere to the whole room.

　　本案为朗诗国际一套平层公寓，设计师在房子原有结构的基础上进一步明确了不同区域的功能性，并稍加改造使其更为合理。本案定位于现代简约风格，以黑白灰为基调，通过黑白灰的深浅变化来展现现代生活的艺术与魅力。房屋的顶部几乎没有修饰，仅仅以简单的石膏线和几盏筒灯来点缀，简约而不失层次感。墙壁也仅仅以现代风格的装饰画点缀，填充空间的视觉效果。客厅配以极具线条感的布艺沙发，电视墙部分设置储物柜和地柜，并在墙壁上安置了隔板，保证了客厅的收纳。会客区和学习区家具选用板式家具，板式家具本身的清新亮丽无需更多的装饰即能带来别样的享受。客厅与书房的过渡区域特意设计了小型的吧台，茗茶品酒如君所需。主卧的墙壁采用了紫色的装饰墙纸，窗帘也配以同色系，紫色带来的高贵、温馨、神秘、典雅的气息弥漫于整个房间。

Youpin Artistic Villa A2 Model House

优品艺墅A2户型样板房

Design Company: Shenzhen Hover House Interior Design Co., Ltd.

设计公司：深圳市世纪雅典居装饰设计工程有限公司

Designer: Nie Jianping

设计师：聂建平

Location: Shenzhen
Area: 96m²
Main Material: grey coated glass marble, carpet, rough surface stone, oil paint, lace curtain

项目地点：深圳
建筑面积：96m²
主要材料：灰色镀膜玻璃、大理石、地毯、毛面石材、油乳胶漆、纱帘

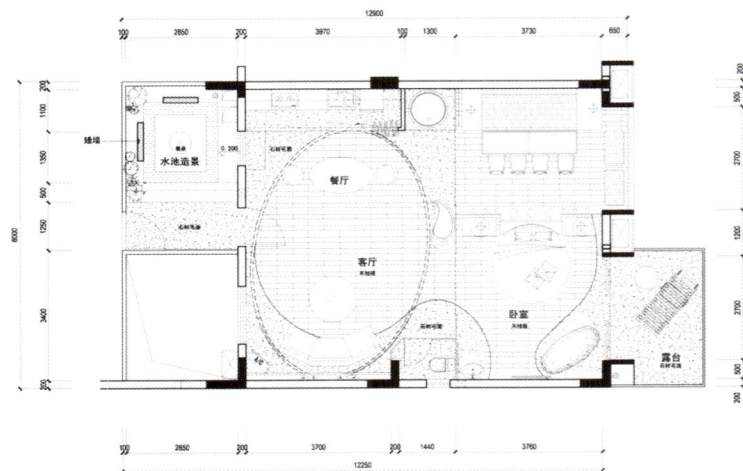

平面图

Based on the original layout, the reconfigured layout just shows a luxury space with warmth and femininity. By removing some redundant dividing walls, the light circulation through the whole flat is allowed. The arch shape of the ceiling and floor creates a sense of flowing, which provides the residents a free limpid peace of mind. Whether precise and elegant or airy and lovely, or pure and soft, or mysterious, all the furnishing in the flat belongs to the owner's favorite.

The rosy background of the living room, and the white lace curtain hang along the ceiling create the change of light and shadow irregularly. The master's bedroom is drowned in an ocean of roses and the super romance means to melt any heart, tough or fragile, and to bury the loss and sadness, conflicts and worries. The lounge room is a best place to share happiness and sadness, or beautiful ideas with a confidant.

　　在原户型通透的基本构架下，重新规划的布局呈现出温暖妩媚的奢华空间。去掉过多的墙隔断，光线可以自由穿透。天花及地面的弧形创造出流动感，置于其中心情如水般自由流淌。室内的一切陈设都是业主喜好的，可以含蓄优雅，也可以轻飘曼妙，可以闪耀着纯洁温柔的光芒，也可以暗藏着神秘，因为在这里只要讨好自己就好。

　　客厅玫瑰红的背景，沿天花弧形的白色丝帘，若有若无间创造空间与光线的变幻。主卧室像是玫瑰花的海洋，极致的浪漫足以柔软每一颗刚毅或脆弱的心，掩埋种种失落与惆怅，矛盾与焦虑。休闲室则是与闺中密友分享快乐忧伤，交流美丽心得的绝佳场所。

Haiyu West Bay Residence

海语西湾

Design Company: Shenzhen Wang Wuping Design Studio

设计公司：深圳五平设计机构

Designer: Wang Wuping

设计师：王五平

Location: Shenzhen
Area: 130m²
Main Material: frosted grey mirror, red oak, wallpaper, polishing brick, grey wood grain marble, mosaic

项目地点：深圳
建筑面积：130m²
主要材料：灰镜磨花、红橡木、墙纸、抛光砖、灰木纹大理石、马赛克

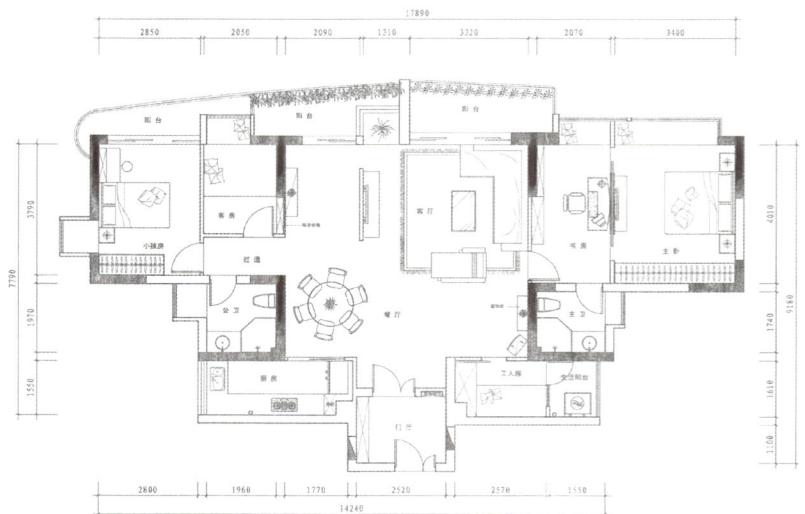

平面图

The house owner knows what she wants in a home style. She likes simple, quiet and chic design with classical black, grey and white colors, which maybe coming from the need of an inner composure after too much luxury and noise. When all the rich classical fantasy is retreating from here, a blank of purity is left by a symphony of black, grey and white.

As for the design method, the blending of black and white is used in simple lines and surfaces to style the whole space. Colors are only limited within the contrast of black and white, without us ng too much colors by means of which further artistic techniques are applied to give a sense of unconventional spatial experience.

Large line-type wallpaper is used as the background of the sofa, which fills the simple space with rhythms. Between the sheer elegance of black and white, flow rich colors, so that the whole space looks ever clean. Meanwhile, the sophisticate decoration of accessories compensates the plainness of monotonous black and white.

　　对于家的格调，业主有着明确的方向，喜欢简洁安静与时尚，并有着经典的黑白灰，也许是源自于内心的一种淡然自若，也或是繁华落尽后的蓦然回首。喜欢就是一种方向，一切关于奢华与古典的畅想在这里业已渐行渐远，留下的便是纯净，是一场黑白灰的时尚交响曲。

　　在设计手法上，本案利用黑白虚实，块面和线条这些最简单有力的语言来设计整个空间，没有运用太多的色彩，而是控制在黑、白两色的对比中，并在空间中运用更艺术化的表现手法来营造其与众不同的空间感受。

　　大幅线形墙纸运用在沙发背景的后面，让素色空间有着律动的节奏。黑白的极致中，丰富的色彩在其中流荡，也使整个空间显得非常干净。同时，在饰品的搭配上也非常巧妙，避免了黑白的单调造成的空洞。

Baoquan Zhuang Residence

宝泉庄

Design Company: Shenzhen Wang Wuping Design Studio

设计公司：深圳五平设计机构

Designer: Wang Wuping

设计师：王五平

Location: Shenzhen
Area: 180m²
Main Material: Dulux paint, archaized brick, laminate flooring, crystal lamp

项目地点：深圳
建筑面积：180m²
主要材料：多乐士乳胶漆、仿古砖、复合地板、水晶灯

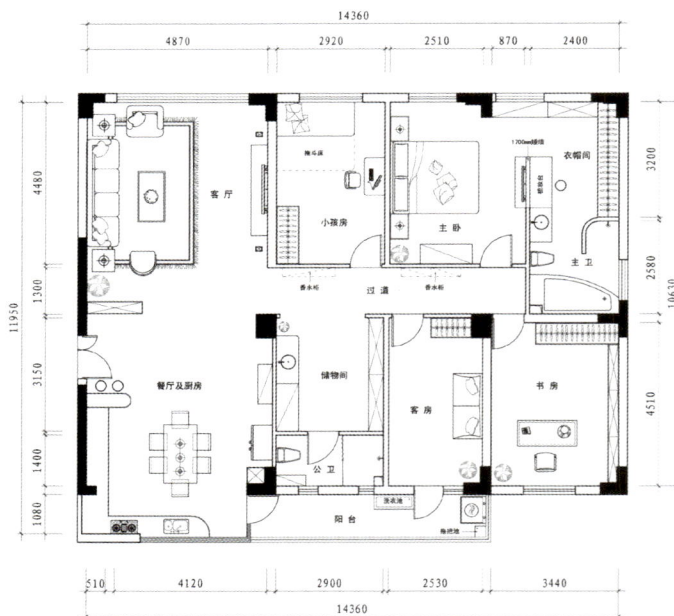

平面图

The blue space is filled with depth and purity, from which rise one's endless imagination. Simple tactile quality, arched lines, and the seashell lights dimly illuminating on the walls of the passageway, all together improvise a blue rhapsody. When wineglasses are held to strike each other, with a little wine evaporating in the air, in the light affect, iridescent colors of happiness would be shaded on the young couple's faces. The tender tactile of soft carpet provides a sense of warmth. If here now came a little music, slow and gentle, it would be a sensuous enjoyment which even wine can not offer.

The moving lines and shapes of this project are originated from an arched line which created from the advantage of the building's original layout. For example, the walls of the living room and dining room stick out to the thickness of a beam, thus being redesigned as an arched shape. And the arch design for the washroom by the passageway, not only divides space but also expands the passageway. As a matter of fact, to design is to look for any existed textures. Then transform it by handling the details with demure but non-implicit techniques.

　　蓝色的空间，充满着深邃和纯净，休闲中有着无限的遐想。纯朴的质感，弧形的动线，过道墙壁上还有着忽明忽暗的几盏贝壳灯，随意间就尽情上演出一曲蓝色的构想。不知何时，杯角间的轻碰，溅起的一沫酒花在空间里漾开，灯光下，斑斓五彩，映在彼此幸福的脸上。柔软的地毯，有一种温暖的感觉，如果再来点舒缓的音乐，相信这是红酒无法营造出的一种精神享受。

　　本案的动线和形体是因现场本身的条件而发展出来的一条弧形的动线，如客厅和餐厅凸出墙面一个梁的厚度，就自然化解为一个弧形。过道旁洗手区用一个弧形门洞设计，既区分了空间，同时又拉宽了过道。其实，设计就像在生活中寻找任何存在的脉络纹理，以一种内敛但不含蓄的手法去处理每一项细节。

Ning

凝

Design Company: D-Hand Design Firm(Fuzhou)

设计公司：鼎汉唐（福州）设计机构

Designer: Chen Mingchen

设计师：陈明晨

Location: Fuzhou
Area: 137m²
Main Material: Portugal cork flooring, rustic brick, leather, wallpaper, laminated flooring, monto paint

项目地点：福州
建筑面积：137m²
主要材料：葡萄牙软木地板、仿古砖、皮革软包、墙纸、金刚板、蒙托漆

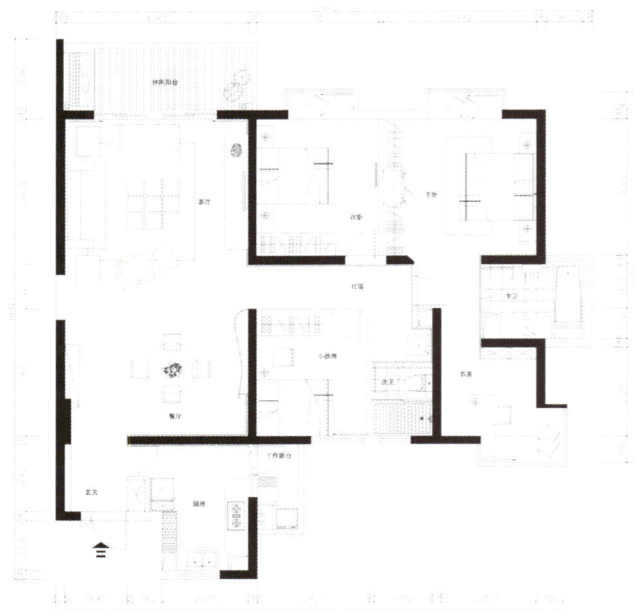

平面图

This project is simple style, the design uses techniques of incision, extension and contrast. And light coffee is the main tone. Through the treatment of large block and surface, the whole space manifests grand and impressive. Material with strong texture interspersed with each other such as cork, laminated flooring, freestone and metal brick, making the space more tasteful. The designer abandons unnecessary complicated ornaments, hidding the doors of different rooms, creating an integrated space.

　　此套住宅为简约风格，设计上使用切割、延伸、对比的手法。色彩以浅咖啡色为主基调。通过大块面及立体构成的处理方式让整体空间彰显大气。材质上采用质感较强的软木、金刚板、毛石及金属砖之间的相互穿插，使空间更具品位。在设计中摒弃多余繁琐的装饰，通过对房间门的隐蔽处理，让整体空间浑然一体。

Gu Yun Fang

古韵坊

Design Company: Shanghai Cayenne Decoration & Design Co., Ltd.

设计公司：上海凯艳装饰设计有限公司

Designer: Zhang Zhisheng

设计师：张质生

Location: Wuxi
Area: 517m²
Photographer: Cheng Xiangyi

项目地点：无锡
建筑面积：517m²
摄影师：程相怡

平面图

This is a simplified European style. Classic European style is luxurious, elegant, harmonious, comfortable and romantic; therefore, it has becoming increasingly popular. European style residence is not only luxurious but also comfortable and romantic. It can bring you and your family endless comfortable feeling through the perfect construction and refined detail processing. Harmony is the highest state of European style. It emphasizes the organization and functions of the space, advocates reasonable technology and respects the nature of materials and is particular about the textures and colors of the materials.

The basement is separated by the glass. It makes the whole space look clean and spacious. A special folding screen behind the tea table makes the area full of cultural connotations. The connection of the inside and outside yards merges perfectly with each other. This is what Chinese garden specializes, that is implication and circuity. The stair has red marble stairways and red sandalwood armrest. If the design used stainless steel as the armrest, it would be too plain. In addition, it would make people feel unsafe. Wooden armrest just mends this defect. The large mirror on the entrance of the 1st floor increases the permeability and extensibility of the space. The glass connects dinning room with kitchen. The repeated European wainscots make the sitting room and dinning room on the same field. The designer paid special attention to the visual effects and layouts. The pure design and color combination make the space clear and concise but still maintains the traditional feeling, which creates this fashion house. On the 2nd floor, the designer uses the simplest decoration to make the space more humanization. The glittering crystal pendant lamp seems to have the magic to string everyone's thoughts to achieve a kind of sympathetic response. An interlined study is added on the 3rd floor in order to create an elegant and quality environment. Sophisticated design conveys a kind of leisure and comfort. The designer not just simply makes the transplantation but to convey the feeling in his own way by creating a residential feeling which combines Chinese and Western features. It has taken great efforts to have all these done. The finished sample is not the same as the original sketch. It has been revised quite a lot and finally had this fashion mansion come out.

Residence is a kind of symbol which emphasizes individual's personalities and tastes. This project has rational space layout, carefully-selected materials and is endowed with human emotions, which indicates people's delicate and quality life in the new era.

本设计采用了简欧的设计风格，欧式风格兼备豪华、优雅、和谐、舒适、浪漫的特点，受到了越来越多的喜爱。欧式的居室有的不只是豪华大气，更多的是惬意和浪漫。通过完美的曲线，精益求精的细节处理，带给家人不尽的舒服触感。实际上和谐是欧式风格的最高境界，重视功能和空间组织。注意发挥结构构成本身的形式美，造型简洁，崇尚合理的工艺，尊重材料的性能，讲究材料的自身质地和色彩的配置效果。

地下一层停车库位与楼梯过道、隔断墙之间采用玻璃分隔断的设计，虚实相连，使得地下室整体空间简洁通透，宽敞明亮；品茶区在独有的屏风衬托下，整个环境大气、稳重、蕴含文化内涵。空间的开敞流通使得室内与室外的庭院彼此间你中有我，我中有你，乃中国园林所讲究的含蓄及迂回。楼梯使用大理石踏步，扶手为红檀木质饰面。若用钢化玻璃做楼梯扶手，虽显得简单现代，但使楼梯整体看上去很单薄，让主人少了分安全的感觉。而木质扶手及宽阔的踏步恰恰弥补了这一不足。一层入口处大块明镜增加了空间的通透性与延伸性。隔断玻璃的使用，使得餐厅和厨房互为一体，欧式壁板造型重复的呈现使得客厅与餐厅连成一片，设计师也特别用心地处理了居高临下的视觉张力与布局。素色设计的色彩搭配，让空间更加简洁、协调而又有传统的韵味，打造出与时尚潮流同步的豪宅。二层设计师在处理空间功能的基础上尽可能以最简洁的点线面装饰，使得空间更合理化。晶莹剔透的水晶吊灯仿佛串联起每个人的思绪，达到一种共鸣。三层为了追求一种豁达、优雅、层次及具有收藏功能的空间环境，增加了一层夹层做书房，细致入微的设计更能表现一种温馨和惬意。设计师并不是简单地将古典移植而是以自己的方式去感知这份悠远情怀，尝试着营造一种具有中西合璧情缘的居室氛围。当初的效果图和修改后的实景图有相当大的不同，设计师最终打造出与时尚潮流同步的豪宅。

居所是个人品位的象征，强调向人文化、艺术化的态势发展。此案整合了空间格局，用料讲究，更置入人的情感归属，体现了新时代人们的精致的生活品质。

Mr. Yang's Mansion, Shalu

沙鹿杨公馆

Design Company: Working Play Design
设计公司：大禾空间设计
Designer: Huang Songxian, Chen Xinhuai
设计师：黄嵩宪、陈欣怀

Location: Taiwan, China
Area: 150m²
Main Material: black marble, paint mandheling veneer, stretch fabric, black mirror

项目地点：中国台湾
建筑面积：150m²
主要材料：黑云石、烤漆曼特宁木皮、绷布、黑镜

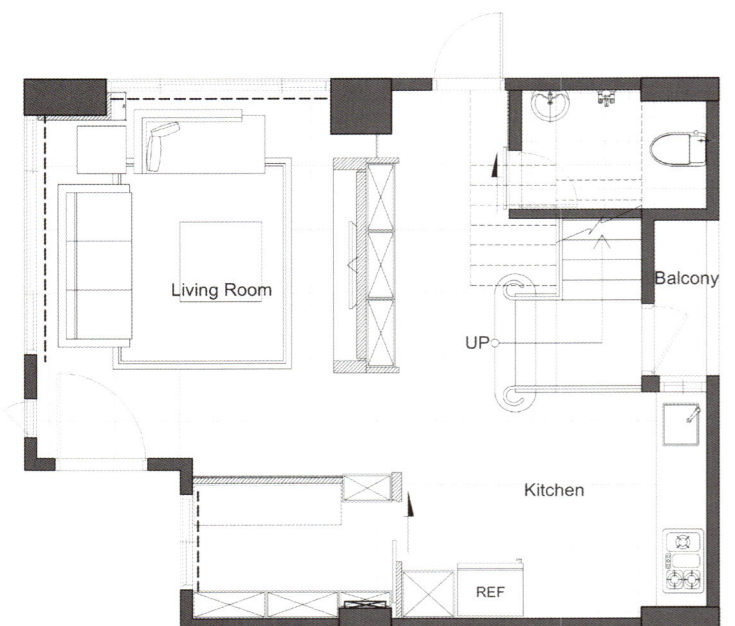

Living Room

Balcony

UP

Kitchen

REF

平面图

The living room which is made from different kinds of stone materials has stable and honorable feeling like a public place. The strong visual contrast conveys a kind of space expansion. You can feel the refined design techniques through the space details. Differing from the color used in public space, the designer applied light colors in this private space. Wooden materials were used as supplement to emphasize the visual impact. The designer successfully created a new living space for the owner through his professional aesthetic knowledge and excellent planning.

In one word, Working Play Design successfully created an aesthetic space which was close to life through good dynamic planning, carefully-selected materials and carefully-calculated proportions. They also hoped their mastery of dynamics and aesthetics as well as the experiences can bring the owner a brand new living enjoyment.

客厅的设计利用不同石材表现出沉稳尊荣的公共空间的感觉。强烈的视觉对比传达出空间的张力，使人们可以从空间的细节中感受到精致的设计手法。不同于公共空间的用色，设计师选用轻浅色系打造私密空间，并辅以木色质感建材强调视觉重心。设计师透过其专业的美学修养及杰出的设计规划成功为业主打造出新的居家氛围。

综观本案，大禾设计透过良好的动线规划、细致的材质运用及比例美学，成功营造出贴近业主生活的艺术空间。设计师们也希望经由其对机能及美学的掌握以技术实现梦想中的居家氛围，为业主带来全新的生活享受。

Chupei

竹北

Design Company: Working Play Design

设计公司：大禾空间设计

Designer: Huang Songxian, Chen Xinhuai

设计师：黄嵩宪、陈欣怀

Location: Taiwan, China
Area: 150m²
Main Material: guanyin stone, geometric cabinet, dark-green paint, teak, black mirror, gloss marble

项目地点：中国台湾
建筑面积：150m²
主要材料：观音石、几何形柜子、深绿漆、柚木、黑色镜面、光面大理石

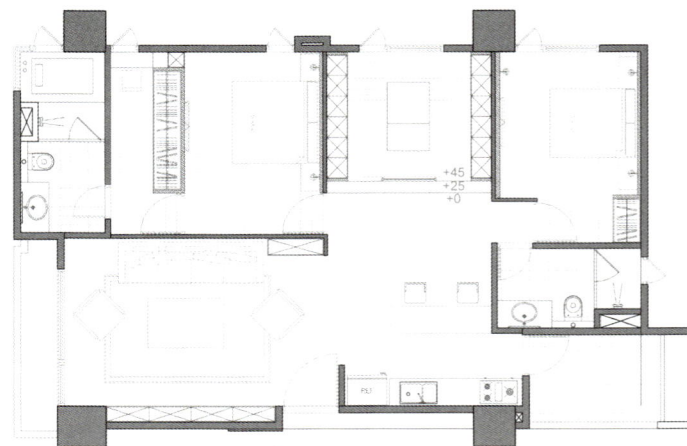

平面图

This apartment is on the thirty floor with three rooms, and one living room and one dining room. The planning is exactly based on what life style the owner needs. The bar located in the center of the space is the major turning point, which links the moving line between the living room and bedrooms. The ceiling is decorated in the concept of layer, which links up the tne living room and kitchen. The walls are decorated with guanyin stone, geometric cabinet, dark-green paint, teak, black mirror, gloss marble, etc.

本案是位于第30层的三房两厅公寓，房子的设计是完全根据房主想要的生活方式来打造的。主要亮点是位于屋子中央连接客厅和卧室的吧台。断层式天花板将起居室和厨房的屋顶连接在一起。墙面的设计采用了观音石、几何形柜子、深绿漆、柚木、黑色镜面与光面大理石等。

Taojin Mountain B1 Type Show Flat

淘金山B1户型样板房

Design Company: Shenzhen Hover House Interior Design Co., Ltd.

设计公司：深圳市世纪雅典居装饰设计工程有限公司

Designer: Nie Jianping

设计师：聂建平

Location: Shenzhen
Area: 300m²
Main Material: off-white marble, wallpaper, teak wood

项目地点：深圳
建筑面积：300m²
主要材料：金龙米黄大理石、柔然墙纸、柚木

平面图

The design combines the traditional Chinese humanity with modern high-quality lifestyle to express the magnificence in the space and to reflect the dignity of the residents, giving a sense of grace and harmony of the contemporary urban nobles who seek for spiritual and quality life.

In order to highlight the magnificence of two-storey space, the limited public area of the original layout was rearranged. Firstly the stairway was moved into the original dining room, and molded into half-spiral arch shape, which has a good connection with the grand character of the living room, dinning room and kitchen. The transparent spacious design improves ventilation and allows more natural daylight. For the second floor, a super high study room was added into the master room as a part of the facilities. The bathroom and the cloakroom were reconfigured to create a sense of luxury which reflect the dignity and temperament of the owner.

The project, extracted from the traditional rectangle and octangle elements, is redesigned and rearranged by using modern decoration techniques. Through a variety of expression on the walls, dividing objects and doors, a coherent quality is extended to reveal the Eastern poetic conception of composure and grace within the luxury. The yellow colors, as one of Easterners' favorite colors, was chosen to be the tone colors which further reflects the sense of luxury. The shining surface of the marble, the gold foil on the furniture and the delicate texture of silk all together interpret the dignity, exuberance and sublime.

设计结合了中国传统人文精神和现代高品质生活方式，力求表现空间的大气，展现出住户的尊贵，营造出现代都市贵族亲切和谐而富有精神追求的优越生活。

为突出双层空间的气派感，将原结构中略显局促的公共区域空间进行了改动。首先将楼梯移至原餐厅处，做成弧形半旋转式成为一景，同时成就了大客厅、大餐厅和大厨房的豪华感，视野通透而开阔，更利于通风与采光。二层增加了一个超高的阳光书房，作为主卧的配套，对主卫及衣帽间进行了重新规划，使主人拥有了豪华的生活配套，尽显主人的尊贵与气度。

装修将传统方形和八角形元素加以提炼，采用现代装饰手法重新进行设计和组合，通过在墙面、隔断、门扇上的不同表现形态，形成一种延续与呼应，使其在华贵中蕴藉儒雅的东方意境。选材和配色以深受东方人喜爱的黄色为基调，充分展现其尊贵感。大理石的光亮和家具的金箔以及丝质布艺的细致肌理，尽显尊贵奢华与气度不凡。

Happiness Coast B1 Type Show Flat

幸福海岸B1户型样板房

Design Company: Shenzhen Hover House Interior Design Co., Ltd.

设计公司：深圳市世纪雅典居装饰设计工程有限公司

Designer: Nie Jianping

设计师：聂建平

Location: Shenzhen
Area: 150m²
Main Material: marble, rose wood, silver foil, brown mirror

项目地点：深圳
建筑面积：**150m²**
主要材料：大理石、花梨木、银箔、茶色镜

平面图

A graceful design is to master the aesthetic image of classical element and to transfer it into modern decoration language. Marble parquet patterns of the floor, archaize anaglyph of the corridor, all of these inspire an exalted temperament. Bold use of color and natural decoration material, creating a sense of enthusiastic and free.

The design lays emphasis on the arc des gn of the furniture, and takes care of the detail, especially the use of soft lines and decorative styles, providing a feeling that it will never out of time. From bed to kitchen, all of these things are full of spanish styles, gorgeous color and various forms, making the entire flat more mystical, steady, and of primitive simplicity.

优雅的设计一向颇能掌握古典元素的美学意象，将之转化为现代装饰语言。地面大理石拼花图案，走道的仿古浮雕，无一不和谐地渲染着尊贵的气质。设计师善于捕捉光线，取材于自然，大胆而自由地运用色彩和造型。

家具注重曲线弧度的造型，工于细节处理，尤其是柔美线条的运用和一些装饰性题材的发挥。其美观程度让人觉得即使经年累月的摆放在家中也不会过时，更显现出贵族的尊贵。从睡床到厨柜摆设，都富有西班牙式的风格，色彩及形象丰富多变，显得更加神秘、内敛、沉稳、厚重、古朴。

Ningbo Yinyi • Lake Garden A2 Type Sample House

宁波银亿·日湖花园A2户型样板房

Design Company: Shenzhen Hover House Interior Design Co., Ltd.

设计公司：深圳市世纪雅典居装饰设计工程有限公司

Designer: Nie Jianping

设计师：聂建平

Location: Shenzhen
Area: 170m²
Main Material: tropical rainforest marble, white microlite stone, oak, Amercian wallpaper, Australian Himars paint

项目地点：深圳
建筑面积：170m²
主要材料：热带雨林大理石、白色微晶石、橡木、美国壁布、澳洲海马斯乳胶漆

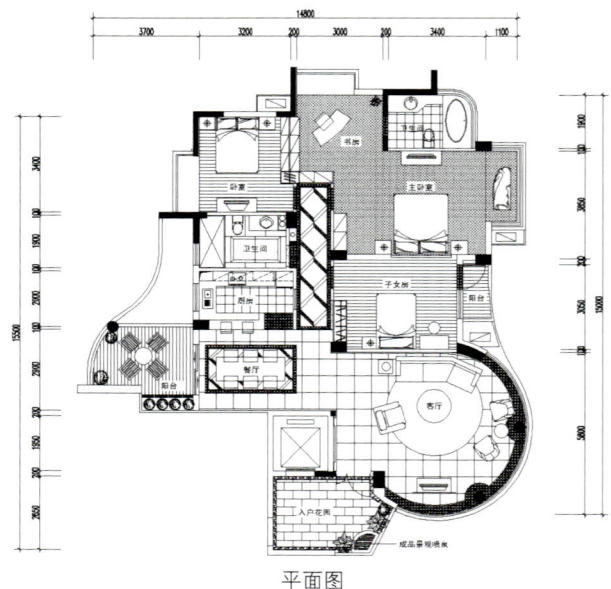

平面图

Designer draws inspiration from Art Deco which featured elegant, simple and delicate. It skillfully avoids the limitation of traditional arrangement and design proportion, meanwhile, it captures and shows the features of Art Deco. Balance, the key of the design, is reflected throughout the space proportion to material and texture.

The natural veins of rainforest marble flooring combine with the bright white microlite stone, making the wallpaper gorgeous. The wrought-iron combines with the dark colored furniture and silvery textile, presenting a refined balance and a vivid effect. The arc-shaped sitting room provides a broad perspective; people could enjoy the beautiful scenery outside the window. The careful selected furnitures make the design looks more elegant. The kitchen is designed as semi-open style. The marble bar counter is used to regale the guests with the wine. The special visual effect of the parquet on porch floor like a road lead to magic kingdom. After a slight arrangement, the master bedroom looks larger than before, highlighting the high quality life of the "urban new nobles".

Designer pays many attentions to the matching between details and the ornaments. Ornament and artworks are visible everywhere in the interior. Such as the curtains from Italy and Belgium, dried flowers from Sri Lanka and the crystal lamps from Egypt with a unique temperament. It would not be a perfect design without all of them.

设计师从典雅，精致而简约的装饰艺术风格中汲取灵感，在捕捉并展现极具风格的装饰艺术要素的同时，以灵活的手法避免一成不变的局限于古典流派的排序及比例。平衡是设计的关键，从空间比例到材料和质地的选择都诠释了这种平衡。

地面富天然纹理的热带雨林大理石充分显示出物料的固有美态，与洁净明亮的白色微晶石结合，衬托出壁纸图案的极致华丽。铁艺雕塑精致高贵，深色的木家具与银色的织物相互融合，展示出精致的平衡，达到鲜明而诱人的效果。圆弧形的会客厅提供宽阔的视野，可欣赏窗外的美丽湖景，精心挑选制定的家具锦上添花，使其彰显尊贵气度。厨房设计成半开放式的，大理石吧用于招待客人畅饮美酒。独特的走廊大理石拼花效果让人生出走过通道便要进入梦幻王国的一般的感觉。主卧室稍加调整后更为宽敞，突显都市新贵的优越生活。

设计师注重家具的细节与饰品的搭配，室内到处可见精致巧妙地装饰品和艺术品，来自意大利和比利时的窗帘、斯里兰卡的干花，还有气质独特的埃及水晶灯，如果缺少了它们，再出色的设计也是不完美的。

International Mansion of China Travel Service Shenzhen Ltd.

深圳中旅国际公馆

Design Company: Gong Decheng Interior Design Firm

设计公司：龚德成室内设计事务所

Designer: Gong Decheng

设计师：龚德成

Location: Shenzhen
Area: 160 m²
Main Material: microlite tile, imported carrara white, golden beige marble, solid wood, imported wallpaper, soft decoration

项目地点：深圳
建筑面积：160m²
主要材料：微晶石地砖、进口大花白大理石、金地米黄大理石、实木线条、进口墙纸、软包

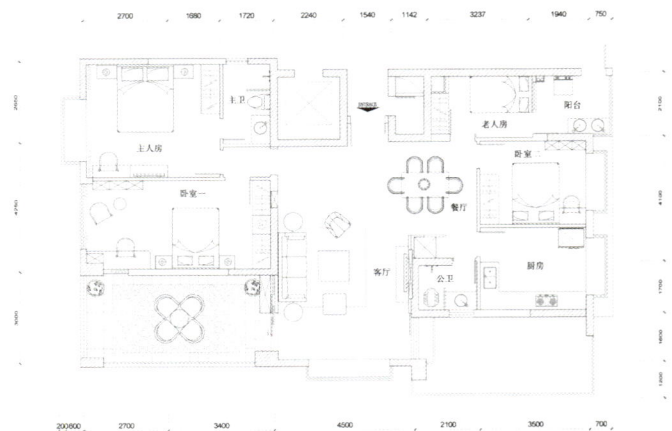

平面图

The concept of this project is low profile luxury. The design team created a high quality stylish residence which demonstrated fashion, humanities and classy living style for the mid-class couple. Design is not merely luxury and brilliance. It should emphasize on connotations. From the selection of the furniture, curtains, wallpaper to the setting of decorative accessories, each detail indicates the understanding of a stylish life. What people need is not just gorgeous appearance, but a quality living style which is called low profile luxury.

本设计理念为"低调的奢华"，设计师为一对中等收入夫妇打造高品位的时尚住家，体现时尚、潮流、人文、品位的家居生活。设计师们不是简单的追求奢侈华丽，而是讲究风格当中的内在品位。从家具选择，窗帘、布艺花纹的搭配，到墙纸图案、色彩的选择以及装饰画的布置，每一个细节都有对品位生活的理解。人们需要的不光是绚丽的外观，更注重的是有品位的生活方式，一种"低调的奢华"。

Civicmoon Villa

信盟别墅

Design Company: Shenyang Grand· Sun Zhigang Space Design Studio

设计公司：沈阳市大展·孙志刚空间设计工作室

Designer: Sun Zhigang, Qian Yu

设计师：孙志刚、钱宇

Location: Shenyang
Area: 800m²
Main Material: marble, ebony, glazed white steel, string curtain

项目地点：沈阳
建筑面积：800m²
主要材料：理石、黑檀木、镜面白钢、线帘

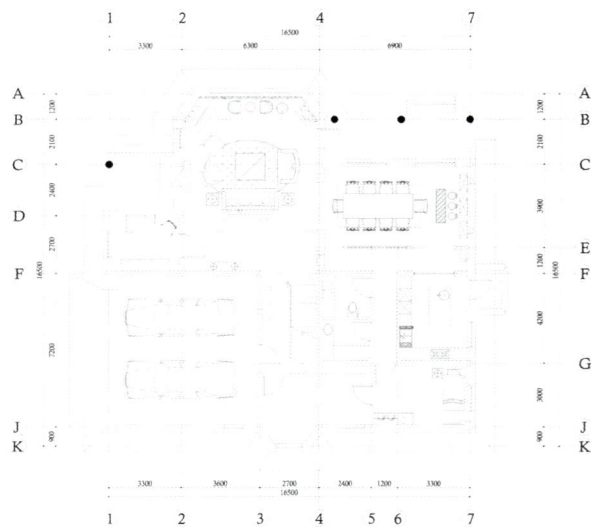

一层平面图

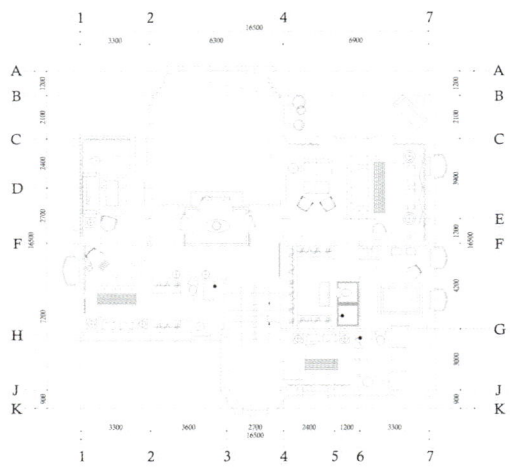

二层平面图

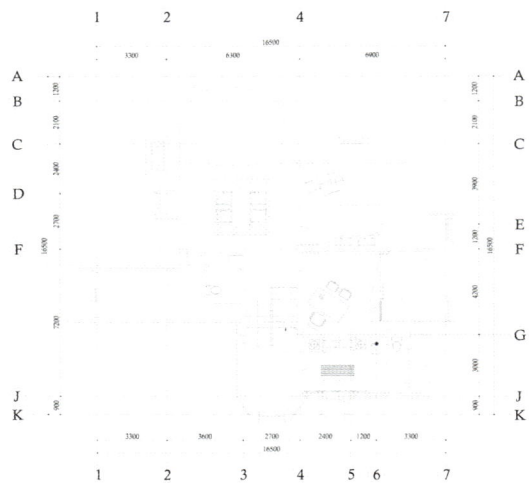

阁楼平面图

Home—a residence without any complexity, is a place where people have deep affection. With the rise of living standard, there comes the era that home has become a place full of inspiration and personalization. Every house owner expects the design of his house can indicate his taste as well as his style.

This is an individual villa which has three floors and one underground floor. Given that the sitting room had very high ceiling, the design decided to make a modern European style. The tedious and complex European symbols were abandoned, but kept the style's elegance and grandness. Light color marble, brilliant crystal chandelier, carefully-selected furniture and accessories, etc. constituted the sedate and noble house. In terms of the master bedroom design, it paid more attention to the lighting design. Monitor control lighting not only made the room has different functions but also added unique flavor to the room. The interesting part was the basement designed as the family salon. It collects the functions of the bar, restaurant and cinema where the owner can treat his friends in this relaxing and comfortable house. Food, drinks, enjoyment, what a great fun!

　　家——居也，一个没有复杂概念，却让人们留恋的地方。随着生活品位的提升，家开始了充满灵感、极富个性的时代。每一位家的主人都期望把自己的家设计成展现个人品位和个性风格的场所。

　　本案例是一处独幢别墅，分为地上三层和地下一层。高挑的客厅增添了空间层次感，于是设计师们利用建筑自身的条件把本次设计定义为现代欧式风格。在设计中舍弃了繁琐复杂的欧式符号却延承了它的大度和精致。客厅里浅色的理石配上华丽的大尺寸水晶吊灯，精心选购的家具和软饰配置的搭配，都使这个家透出一种稳重华贵的气质。在主人卧室的设计中着重于灯光的把握。分路控制灯光，不但使房间在使用功能上有不同的变换，同时为房间增添了不同的气氛。在本次设计中还有一处令人喜悦的地方，就是安排在地下室的家庭沙龙，它把酒吧、餐馆、影视厅的所有功能聚集一身，使朋友在这自由、轻松，又充满家的温馨的房子里尽情畅饮，真是不亦乐乎！

Slow Living Love Villa

慢活爱的别墅

Design Company: Lunarian Creative
设计公司：宽月空间设计
Designer: Billy Wu, Shemy Ta
设计师：吴奉文、戴绮芬

Location: Taiwan, China
Area: 120m²
Main Material: terrazzo, smooth pebble, porcelain tiles, slate, bamboo flooring, sandblast wood veneer, fir wood board, calcium silicate board, tempered glass, natural green paint, stainless steel+fluorocarbon baking varnish, low-e, energy-saving glass, southern pine

项目地点：中国台湾
建筑面积：120m²
主要材料：磨石子、抿石子、石英砖、板岩、竹片地板、喷砂木贴皮、杉木板、硅酸钙板、强化玻璃、天然环保涂料、不锈钢＋氟碳烤漆、低辐射镀膜节能玻璃、南方松

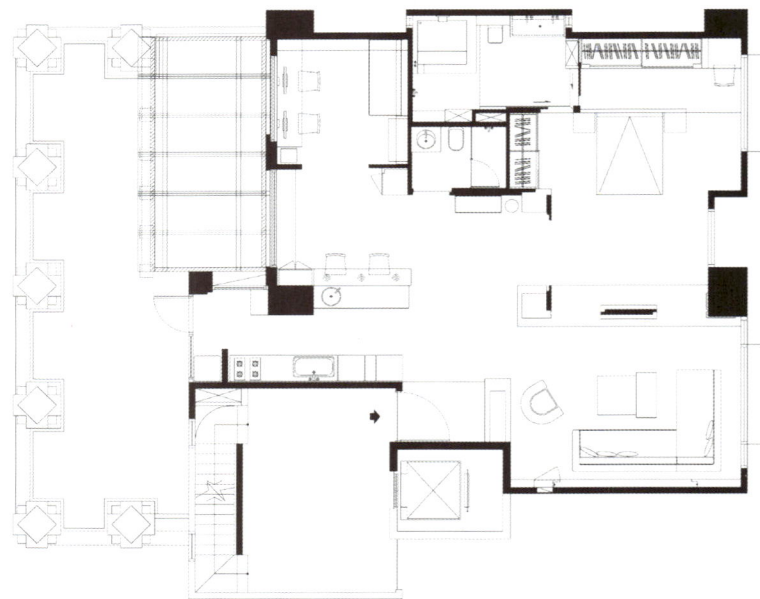

平面图

This project is to reconfigure space based on the spacious visual penetration. At the corners of the seemingly detached retreat dwelling hide convenient storage cabinets. The sophisticated arrangement of visual and function well explains the combination of craftsmanship and aesthetics. Natural visual sense for the overall design and ecology sustaining have been taken in consideration in the use of wooden materials.

The wall is made of massive polished stones and little pebbles. And the woodworks are made of rough wood veneer toned with the colors of earth which makes it appear unostentatiously elegant. All the lighting products through the dwelling are energy-saving light bulbs and tubes indirectly having a play on light and shadow, and together with the lighting decoration creating nice atmosphere. The plants placed indoor have a good connection with the outdoor space, which not only forms a picturesque view but also makes a little contribution to the prevention of increasing global warmth. The decoration sold on market and a variety of creative DIY ideas have been used to transform the original space into a unique retreat heaven available to all seasons.

Entering the vestibule, stepping onto the polished stones and feeling the simple texture, one can easily spot the different way of hiding the vestibule to the ordinary. A row of elegant white linear-curtain lamp shade separates the living room from other spaces. The living room is a place for getting together watching TV and enjoying music. The hanging design of placing the TV saves space for the living room. And even if the room has different multi-channel and bi-channel audio equipment, and various shapes and sizes of audio and optic equipments, the advanced planning and precise design will hide well all of the electric lines and place the equipments within the wooden pillar. The

woodwork behind the sofa is not only for covering electric equipments but also is for storage. The green succulent plants naturally fit into the background of woodworks. The comfortable L-shaped sofa in the living room companied with the "Π" shaped tea table and its auxiliary stool replacing the traditional living room bench, makes the functional space more enjoyable.

本案的设计主轴在于以开阔的视觉穿透感来安排空间的配置，看似不食人间烟火的度假别馆，实际上却将机能和收纳藏匿于不易被发现的四周角落。视觉与机能的巧妙安排，达到工艺与美学的完美结合！在布置材质选用方面，不单是整体视觉以自然感呈现，并严选环保素材来制作，不仅美观更为健康把关！

墙面以大量磨石子、抿石子为主题，而木作也以粗木皮来制做，配色用了大量的大地色系，呈现出朴素优雅却低调的质感。灯光方面，全屋内外使用省电灯泡和灯管，可节约能源。另外大量利用间接照明的方式来玩光影游戏，精心选配的灯饰，也是营造氛围的好帮手。大量的植物和室内外空间相辅相成，除了构成如同大自然般的美丽图画，同时也为防止地球暖化尽一份心力！最后装饰期间，加入各种自助改造的想法。即使是市售的装饰品，也能变成匠心独具的自我风格。如此，打造出这个全年无休的度假天堂。

进入玄关，脚下即可感受到磨石子朴素的质感，不同于一般玄关的完整遮蔽，本案以一排素雅的白色线帘灯罩作为和客厅的区隔。客厅可热闹地看电影，同时也能宁静地听音乐。悬挂在木作墙面上的电视节省了放置的空间，即使客厅拥有多声道和二声道不同的音响设备以及大大小小的不同外型的声光机器，在事先规划和精密的设计下，管线全都藏于无形，并将机器收纳在精心设计且通风的木作长柱里。沙发后的木作除了遮掩电箱之外，也是收纳的好帮手。此外更将多种植物的绿意融入木作背景中。客厅选用舒适的L形沙发，以木作做了"Π"形桌结合脚凳取代了传统客厅的长桌，如此可更轻松享用客厅的机能空间。

Secret Alley

桃园洪邸

Design Company: XYI design consulting Co., LTD

设计公司：隐巷设计顾问有限公司

Designer: Mac Huang, Carrie Meng

设计师：黄士华、孟羿彣

Location: Taiwan, China
Area: 160m²
Main Material: phoenix tree wood, KD walnut wood, paint, glass, lime stone, South Africa light black marble, grey mirror, piano lacquer, kronotex wood flooring.

项目地点：中国台湾
建筑面积：160m²
主要材料：梧桐实木、KD 胡桃实木、烤漆玻璃、石灰石、南非浅黑烧面、灰镜、钢琴烤漆、金钢柏林木地板

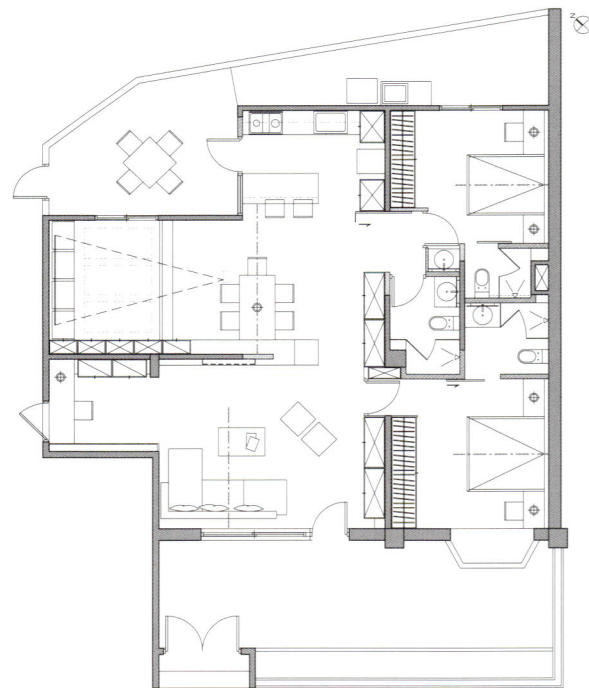

平面图

The material select on of this case is focusing on the original texture, abandoning overmuch adornment material, retaining the sober toned space. The character of the space will be decided by the homeowner. The living space needs the sense of stability and quiet. Moist wood matches with beige marble, making people feel relaxed. The sandwood scent and the bamboo flooring make people feel ease, and the heath concept arises spontaneously.

　　此案的材料着重在原始质感上，舍弃过多的装饰材料，保留色调素静的空间，空间形象特质交由屋主发挥。人居空间需要稳定感与宁静感，温润的木头搭配米色火烧面大理石，使人精神放松。檀香味和竹制实木地板使人通体放松，养生的概念油然而生。

Happiness Coast A3 Type Sample House

幸福海岸A3户型样板房

Design Company: Shenzhen Hover House Interior Design Co., Ltd.

设计公司：深圳市世纪雅典居装饰设计工程有限公司

Designer: Nie Jianping

设计师：聂建平

Location: Shenzhen
Area: 300m²
Main Material: bianco botticino marble, black marquina marble, white crystallite stone, mural tile, mosaic, ebony, ebony flooring, wallpaper, black slate

项目地点：深圳
建筑面积：300m²
主要材料：白砂米黄大理石、黑白根大理石、白色微晶石、墙砖、马赛克、黑檀、黑檀木地板、墙纸、黑石板

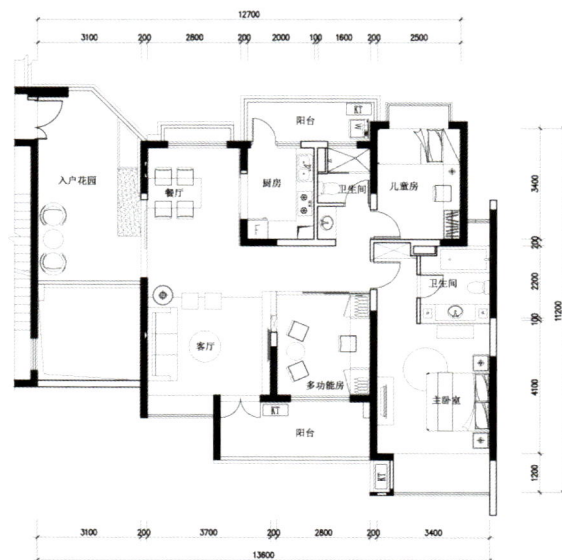

平面图

This project is toned with black, applied with the interweaving of lines and surfaces, and combined with Art-styled pattern, unique lighting and decoration to create high-class residential atmosphere of comfort.

The walls of the living room, themed by black and white wallpaper and hung with sun-shape decorations, sparkle with energy. The edgy black chandelier hung from the ceiling and the black lamp beside the sofa perfectly reveal the theme of Art Deco style. The cozy cushion placed on the window sill is a big focus point which not only takes clever advantage of the space but also provides a pleasant romantic sense. The design of the children's bedroom fully reflects the designer's consideration of humanization. The rearrangement of the furniture makes good use of the space maintaining the comfortable quality meanwhile. Personality and taste is the highlight of this project—it is not necessary to be the most luxury, but it absolutely should be most suited to themselves.

在本案中，设计师以黑色为基调，充分运用线与面的穿插，结合艺术风格的图案、纹样，配以独具风格的灯具、饰品，营造出高品位的舒适的居家氛围。

客厅主题墙面是黑白条纹的壁纸，悬挂着太阳状造型饰物，给人朝气蓬勃的感觉。天花结合极具设计感的黑色吊灯和沙发旁的台灯，将整个设计的现代装饰艺术主题风格完美地烘托出来。在飘窗上放置舒适的坐垫成为一大亮点，不仅合理利用了空间，还营造出浪漫、惬意的感觉。儿童房的设计充分体现了设计师的人性化思维，组合家具使空间得到合理利用，而又不失舒适。本案设计师着重强调个性及品位——不必是最奢华的，但一定是适合自己的。

REFORMA RESIDENCE

雷福马住宅

Design Company: Pascal Architectural Design Firm

设计公司：帕斯卡尔建筑设计事务所

Designer: Carlos Pascal, Gerard Pascal

设计师：卡洛斯·帕斯卡尔、热拉尔·帕斯卡尔

Location: Av. de la Reforma, Ciudad de México
Area: 405m²
Photographer: Sófocles Hernández

项目地点：墨西哥城阿维尼达雷福马
建筑面积：405m²
摄影：索福克勒斯·埃尔南德斯

平面图

Due to the facade and structure complexity of the building, the remodeling apartment design is based on a logical existing elements union. Instead of hiding them, the design creates spacious and illuminated areas with different perspective views in each one. This traces generated angulated views and, therefore independent space that do no need to be closed by walls or doors; these are the aspects that portrays the house at first sight. The core element upon which the space was designed is a laminated metal sheet that runs through the bedrooms hallway, the guest washroom, the elevator, and goes around up to the dining room becoming an important part of it too. A huge bookcase, that do not reach the ceiling, follows the same rules though in a perpendicular orientation. These are the only two elements that defines the public space limits in the house. Another important aspect of the design are the minimun required circulation areas; corridors become part of the adjacent zones. This is how getting around the house becomes very comfortable and practical having more than one route to choose. The functional design and the visually appealing and habitable perspectives, are combined with a rich materials and colors palette: black slate, metal, wood and glass. All of them were selected to bring out the best of each one; the textures, shades and the highlights are unique elements that work together creating a strategic combination. The black slate used in the hallway, the entrance foyer and the dinning room floors are also in the bathrooms area but in combination with walls covered with two tonality Venetian tile. The guest washroom walls are lined with cow black leather emphasizing the Venetian mirror luminosity and the granite sink.

　　由于楼宇大外结构复杂，设计师在改造此项目时，保留、融合了原有元素，重整出宽敞明亮的居住空间。从每一个区块观察，各自形成不同的透视角度。在视觉上各区域空间自成一统，因此也免除了门与墙的累赘使用。这些特点是此案给人视觉上的第一印象。此案的焦点在于以一复合金属板贯通卧室走廊、客卫、电梯并围绕餐厅成为其一部分，同时利用一组不靠顶的大书柜以相似方式构成垂直角度连接各区块。以上也是此案隔断室内公共区域空间的两种方法。此项室内设计的另一特色是将流动区域减少到最小限度，模糊了走廊与毗邻区域的分界线，无形中增添了另外的流动通道，大大提升了空间的舒适性及功能实用性。黑板岩、金属、木材与玻璃等多种用料与丰富的色彩搭配完美诠释了集功能性、宜居性及视觉震撼力为一体的室内设计。设计师对材料精挑细选，将各自的优点发挥得淋漓尽至，运用纹理、光影、对比等技法共同组合打造出独一无二的空间。黑板岩大量运用于玄关、门厅及餐厅地板处。浴室中的黑板岩与双色调威尼斯瓷砖相搭配。客卫墙壁则镶以线形黑色牛皮以突出威尼斯镜子的明度与花岗岩盥洗盆的质感。

PROLONGACIÓN BOSQUES

森林延续

Design Company: Pascal Architectural Design Firm

设计公司：帕斯卡尔建筑设计事务所

Designer: Carlos Pascal, Gerard Pascal

设计师：卡洛斯·帕斯卡尔、热拉尔·帕斯卡尔

Location: Av. de la Reforma, Ciudad de México
Area: 450m²
Photographer: Sófocles Hernández

项目地点：墨西哥城拉斯洛马斯森林
建筑面积：450m²
摄影：索福克勒斯·埃尔南德斯

平面图

Located in one of the most exclusive areas of Mexico City this apartment represents good taste a a way of living. The apartment is distributed around a central core of services and it is unified as a whole around these. In all areas the materials used Palissandre wood from India, which has a very distinctive grain, and Italian silver travertine marble that creates a contrast of color and texture. Most of the furniture and lighting was designed by Pascal Architectural Design Firm to generate visual language unity; classic and contemporary art collection is on display.

此公寓位于墨西哥城最顶级的区域，象征着优质品位引领的生活方式。公寓设计以一系列核心服务设施为中心展开布置，以统一风格贯通融合。各区域所使用的紫檀木均来自印度。带有十分独特的纹理，而意大利钙华大理石的运用则在材料质地与颜色上形成了对比。大部分的家具与照明均由帕斯卡尔建筑设计事务所设计布置，配以经典与现代的艺术陈列品，营造出视觉上的统一效果。

Yonghe Ye's Mansion

永和叶邸

Design Company: Cain Lane Design Consulting Co., Ltd
设计公司：隐巷设计顾问有限公司
Designer: Mac Huang, Carrie Meng
设计师：黄士华、孟羿彣

Location: Taiwan, China
Area: 85m²
Main Material: oak veneer, paint glass, grey glass, white paint, wood grain PVC flooring, special paint

项目地点：中国台湾
建筑面积：85m²
主要材料：橡木实木皮、烤漆玻璃、灰色玻璃、白色烤漆、木纹聚氯乙烯地坪、特殊漆

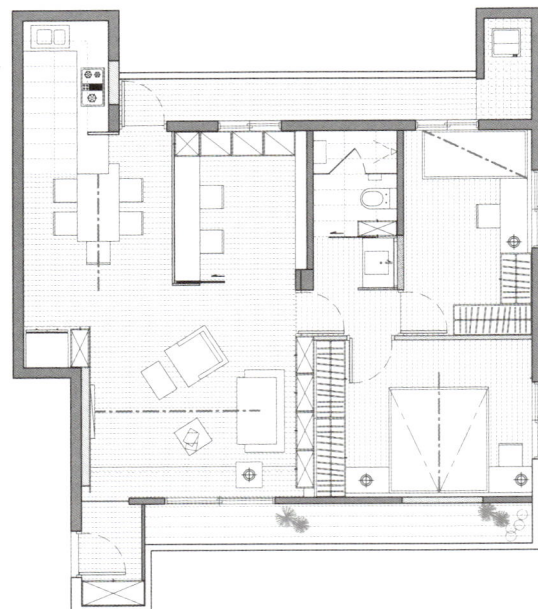

平面图

This case focuses on showing indoor layout, configuration and space transformation suitable for living needs. Through the design devices, the design reduces tedious moving line. Black ,white and grey are the primary color matching for the interior, combining the texture of the selected material and the refined quality of the accessories, perfectly showing the elegant humanistic manner, pure and fresh. In the limited interior space, through the design, each function zone is linked together, maximizing the space effect. Through the alternative of lighting, scale and material, the space function is strengthened, meanwhile the interactive range is hightlighted, which make this 30-year-old house sparkle with new meaningful theme. The classic hues of black, white and grey integrating with the neat and simple style, emphasizes the quiet of the space. The ower's fast-paced life especially defines home as a comfortable carefree harbor. The whole design tone creates the space with spatial extent and brightness, at the same time it also considers about the psychology element, hightligting the designers" careful thought. The unique union of the ceiling and the lamps interprets the new design concept neatly and clearly. The furniture and adornments in the dining room and study room, besides reflecting the owner's living habits, help the owner in avoiding the sensory untidy feeling.

　　本案着重展现室内格局、配置及适应生活需求的空间转换，通过设计手段，减少冗长动线。室内以黑、白、灰等低调色系为首要配色，结合选材的质感与配饰的细致，完美展现了优雅的人文气度，脱凡超俗。在有限的室内空间中，透过设计串联每个区域的机能属性，让空间发挥到最大效果。通过光线、比例、材质的重复交迭，突出互动范围的同时，也使空间机能增强，让这间 30 年的老宅，闪动出新的主题意义。黑、白、灰的经典色调，结合利落、简约的设计风格，强调了空间的静谧感。屋主有着快节奏的生活步调，定义家是令人舒适居住的悠然港湾。设计选择的整体色调，在创造空间范围跟亮度的同时，还考虑到心理学元素"减压"，突显出设计师的细致考虑。吊顶与灯光的巧妙结合，干净利落地诠释了新的设计理念。餐厅、书房的家具配饰，在体现了主人生活习惯之余，也为主人排除掉室内凌乱的感觉。

本书参编人员（排名不分先后）：

邹志雄	戴　勇	李益中	黄士华	冯　羽
邱春瑞	刘　鹏	赵牧桓	王春添	邱宜平
黄怀德	方　峻	赵力行	琚　宾	洪德成
刘晓都	孟　岩	王　辉	刘卫军	韩　松
张质生	黄鹏霖	林宪政	李　泷	王晓冬
卢积灿	王　创	王　宇	张喜娜	杨思雁
肖湘桂	罗　斌	陈桂娇	洪　辉	丁小娟
项国媚	陈　龙	黄　丹	王　佳	郑思瑜
李　莹	李　忍	李　静	鲍　威	刘莉莉
彭　莹	罗　方	胥群燕	李文珠	雷　云
洪　霖	江　勇	马志远	王佳丽	张　蕾
陈　静	沈　婷	胡晓意	肖珍华	陈思媛
杨创创	冯　韬	李继旺	周　晶	曹小云
李姣姣	李　媛	申　琦	肖　广	李　强
梁义芳	王　伟	林宗明	梁春苗	黎文君